SpringerBriefs in Speech Technology

Studies in Speech Signal Processing, Natural Language Understanding, and Machine Learning

Series Editor

Amy Neustein, Fort Lee, NJ, USA

SpringerBriefs present concise summaries of cutting-edge research and practical applications across a wide spectrum of fields. Featuring compact volumes of 50 to 125 pages, the series covers a range of content from professional to academic. Typical topics might include:

- A timely report of state-of-the-art analytical techniques
- A bridge between new research results, as published in journal articles, and a contextual literature review
- A snapshot of a hot or emerging topic
- An in-depth case study or clinical example
- A presentation of core concepts that students must understand in order to make independent contributions

Briefs are characterized by fast, global electronic dissemination, standard publishing contracts, standardized manuscript preparation and formatting guidelines, and expedited production schedules.

The goal of the **SpringerBriefs in Speech Technology** series is to serve as an important reference guide for speech developers, system designers, speech engineers and other professionals in academia, government and the private sector. To accomplish this task, the series will showcase the latest findings in speech technology, ranging from a comparative analysis of contemporary methods of speech parameterization to recent advances in commercial deployment of spoken dialog systems.

More information about this series at http://www.springer.com/series/10043

Manjunath K. E.

Multilingual Phone Recognition in Indian Languages

 Springer

Manjunath K. E.
U R Rao Satellite Centre
Indian Space Research Organisation
Old Airport Road, Bengaluru
Karnataka, India

ISSN 2191-737X   ISSN 2191-7388   (electronic)
SpringerBriefs in Speech Technology
ISBN 978-3-030-80740-5   ISBN 978-3-030-80741-2   (eBook)
https://doi.org/10.1007/978-3-030-80741-2

This Springer imprint is published by the registered company Springer Nature Switzerland AG
The registered company address is: Gewerbestrasse 11, 6330 Cham, Switzerland

Preface

India is a land of many languages, among them 122 languages are spoken by at least 10,000 people each, with 22 of them constitutionally recognised. Several of the Indian languages do not have sufficient labelled data to develop a separate phone recogniser for themselves. This necessitates an investigation into alternative ways of performing phone recognition, such as multilingual phone recognition. A Multilingual Phone Recognition System (Multi-PRS) is a language-independent, universal Phone Recognition System (PRS) that can recognise the phonetic units present in a speech utterance independent of the language of the speech utterance.

In this book, various aspects of multilingual phone recognition such as development, analysis, performance improvement, and applications of Multi-PRSs are studied for *six* Indian languages – Kannada (KN), Telugu (TE), Bengali (BN), Odia (OD), Urdu (UR), and Assamese (AS). Among the *six* Indian languages considered, Chaps. 3, 4, and 5 use only *four* languages (KN, TE, BN, OD), while the Chap. 6 uses all the *six* languages. The International Phonetic Alphabets (IPA) based transcription is used for deriving a *common multilingual phone-set* by grouping the acoustically similar phonetic units from multiple languages. Both *Gaussian Mixture Model (GMM)-Hidden Markov Models (HMM)* and *Deep Neural Network (DNN)-HMMs* are explored for training the Multi-PRSs using Mel-frequency Cepstral Coefficients (MFCCs) as features under *context independent* and *context dependent* settings. The behaviour of Multi-PRSs across *two* language families namely – Dravidian and Indo-Aryan – is studied and analysed by developing separate Multi-PRSs for Dravidian and Indo-Aryan language families. The performance of Multi-PRSs is analysed and compared with that of the Monolingual Phone Recognition Systems (Mono-PRS).

Articulatory Features (AFs) are explored to improve the performance of Multi-PRSs. The AFs for five AF groups – place, manner, roundness, frontness, and height – are predicted from the MFCCs using DNNs. The oracle AFs, which are derived from the ground truth IPA transcriptions, are used to set the best performance realizable by the predicted AFs. The performance of predicted and oracle AFs are compared. In addition to the AFs, the phone posteriors are explored to further boost the performance of Multi-PRS. Multitask Learning (MTL) is explored to improve the prediction accuracy of AFs and thereby reducing the Phone Error Rate (PER) of Multi-PRS. Fusion of AFs is done using two approaches: (i) lattice rescoring approach and (ii) AFs as tandem features. It is found that the oracle AFs by feature fusion with MFCCs offer a remarkably low target PER of 10.4%, which is 24.7% absolute reduction compared to baseline Multi-PRS with MFCCs alone. The fusion of phone posteriors and the AFs derived from the MTL yields the best performance. The best performing system using predicted AFs has shown reduction of 3.2% in absolute PER (i.e. 9.1% reduction in relative PER) compared to baseline Multi-PRS.

Applications of multilingual phone recognition in code-switched and non-code-switched scenarios are discussed. Two different approaches for multilingual phone recognition using code-switched and non-code-switched test sets are compared and evaluated. First approach is a front-end Language Identification (LID) system followed by monolingual phone recognisers (LID-Mono) trained individually on each of the languages present in multilingual dataset, while the second approach uses a *common multilingual phone-set* without requiring a front-end LID based switching. Bilingual code-switching experiments are conducted using the code-switched test sets of Kannada and Urdu languages. The state-of-the-art i-vectors are used to perform LID in first approach. It is found that the performance of *common multilingual phone-set* based approach is superior compared to more conventional LID-Mono approach in both non-code-switched and code-switched scenarios. The performance of LID-Mono approach heavily depends on the accuracy of the LID system, and the LID errors cannot be recovered. However, the *common multilingual phone-set* based approach by virtue of not having to do a front-end LID switching and designed based on the common multilingual phone-set derived from several languages is not constrained by the accuracy of the LID system, and hence performs effectively on non-code-switched and code-switched speech, offering low PERs than the LID-Mono system.

This book is mainly intended for researchers working in the area of multilingual speech recognition. This book will be useful for the young researchers who want to pursue research in speech processing with an emphasis on multilingual speech recognition. Hence, this may be recommended as a text or reference book for the postgraduate-level advanced speech processing course. The book has been organised as follows:

Chapter 1 introduces basic concepts of multilingual speech recognition. The multilingual AFs and code-switched speech recognition are briefly introduced. Chapter 2 describes the prior work on multilingual speech recognition systems with primary focus on multilingual AFs and code-switched speech recognition. Chapter 3 describes the development, evaluation, and analysis of Multi-PRS for four Indian languages. Chapter 4 discusses the proposed approaches to derive the multilingual AFs from spectral features. Chapter 5 discusses the use of predicted multilingual AFs to improve the performance of Multi-PRS. Chapter 6 describes the applications of multilingual phone recognition in code-switched and non-code-switched scenarios. Two approaches for multilingual phone recognition are compared using code-switched and non-code-switched test sets. Chapter 7 provides a brief summary and conclusion of the book with a glimpse towards the scope for possible future work.

I am grateful to my PhD supervisors Prof. V. Ramasubramanian and Prof. Dinesh Babu Jayagopi at the International Institute of Information Technology, Bangalore (IIITB), for their constant support, guidance, and encouragement to carry out this work. This book is based on my doctoral thesis work. I am also grateful to my MS supervisor Prof. K. Sreenivasa Rao at IIT Kharagpur for providing the speech corpora to carry out this work. I thank all the professors and research scholars of IIITB who have helped me to carry out this work. Special thanks to ISRO management and to my colleagues at URSC, ISRO for their cooperation and encouragement during the course of editing and publishing this book. Last but not the least, I am grateful to my parents, my in-laws, my wife, and my daughter for their constant support and encouragement. Finally, I thank all my friends and well-wishers.

Bengaluru, India   Dr. Manjunath K. E.

Contents

Acronyms

AF	Articulatory Feature
AF-PER	AF-Prediction Error Rate
AF-Predictor	Articulatory Feature Predictor
AF-Tandem	Combination of AFs as Tandem features
AS	Assamese
ASR	Automatic Speech Recognition
BN	Bengali
CD	Context-Dependent
CI	Context-Independent
DNN	Deep Neural Network
DP	Dynamic Programming
FFNN	Feed-Forward Neural Network
GMM	Gaussian Mixture Model
GMM-UBM	Gaussian Mixture Universal Background Model
HL	Hidden Layer
HMM	Hidden Markov Model
Hz	Hertz
IPA	International Phonetic Alphabet
ISA	Intrinsic Spectral Analysis
KN	Kannada
LFV	Language Feature Vector
LID	Language Identification
LID-Mono	LID-switched Monolingual Approach
LRA	Lattice Rescoring Approach
LVCSR	Large Vocabulary Continuous Speech Recognition
MFCC	Mel-frequency Cepstral Coefficient
MLP	Multi-layer Perceptron
Mono-PRS	Monolingual Phone Recognition System
ms	Millisecond
MSE	Mean Squared Error
MTL	Multitask Learning

Multi-PRS	Multilingual Phone Recognition System
OD	Odia
PER	Phone Error Rate
PP	Phone Posterior
PRS	Phone Recognition System
Quadri-PRS	Quadrilingual PRS
seq2seq	Sequence-to-Sequence
SVM	Support Vector Machine
TE	Telugu
UR	Urdu
WER	Word Error Rate

Chapter 1
Introduction

1.1 Multilingual Phone Recognition

India being a country of many multilingual societies, with at least three languages being spoken in most of the major cities, the need for developing multilingual speech recognition systems is becoming increasingly important. Some studies claim that more than half the world's population is multilingual [1]. This claim would be more appropriate for a multilingual country like India, which is a home to a large number of languages. India has 122 major languages [2] among them 22 are constitutionally recognized [3]. Since most of the Indian languages belong to one of the four language families—Indo-Aryan, Dravidian, Austro-Asian, and Sino-Tibetan [3]—and the languages within a language family have large linguistic similarities in their acoustic phonetics, articulatory phonetics, grammatical structures, and vocabulary patterns [4], the development of multilingual speech recognition systems is very appropriate in the context of Indian languages. Hence, it is essential to investigate the development of multilingual speech recognition systems using Indian languages.

Automatic Speech Recognition (ASR) is used for decoding the message conveyed in the speech signal to text (stream of words). A full fledged ASR system consists of several components such as Phone Recognition System (PRS), phoneme-to-grapheme conversions, and language models. The PRS, that is used for decoding the phonetic units present in the speech signal, is one of the central techniques in ASR. A PRS can be either language dependent or language independent in nature depending on the data with which it is trained. Figure 1.1 shows the block diagram of a PRS. It accepts speech signal as input and decodes it into sequence of phones at its output.

The *Mono-PRS* is trained using the data of a single language and hence it is language dependent in nature, whereas the data from multiple languages is pooled together to develop a Multi-PRS; this makes it to be language independent in nature. Multi-PRS is a language independent, universal PRS that could recognize the phonetic units present in a given speech utterance independent of the language of the

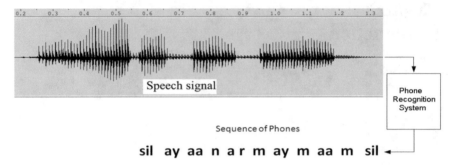

Fig. 1.1 Block diagram of a Phone Recognition System

speech utterance. The problem with the Mono-PRS (i.e. language dependent PRS) is that it requires a separate PRS to be developed for each language, which might not be feasible in reality due to non-availability of training data in all the languages. On the other hand, a Multi-PRS can be used to develop phone recognizers for under-resourced (or low-resourced) languages [5]. Multilingual ASR has its applications in machine translation, speech-to-speech systems [6, 7], language adaptation [8], language recognition [9] and code-switching [10–12]. In this book, various aspects of multilingual phone recognition such as development, analysis, performance improvement, and applications are studied using six Indian languages—Kannada, Telugu, Bengali, Odia, Urdu, and Assamese.

1.2 Articulatory Features for Multilingual Phone Recognition

"One of the important direction in multilingual speech recognition is the use of AFs, given that their production basis serves as a common feature set across languages. AFs represent the positioning and movements of articulators during the production of a sound unit. They are related to the speech production mechanisms and provide a more compact representation of the speech independent of the language in which it is produced. AFs represent a higher degree of *invariance* and hence it will be very appropriate to use them in multilingual tasks so that the acoustic-phonetic variability across languages [13] can be captured.

The articulators such as lips, teeth, tongue, alveolar ridge, hard palate, velum, and glottis are involved in the speech production. AFs change from one sound unit to another, and contain lexical and phonetic information. The IPA chart is designed based on the speech production characteristics (i.e. AFs representation) of each sound unit [14]. AFs can be broadly classified into five groups namely—(i) Place, (ii) Manner, (iii) Roundness, (iv) Frontness, and (v) Height. The place and manner AF groups capture the characteristics of consonants, while the roundness, frontness, and height AF groups capture the characteristics of vowels. The significance of having five AF groups to capture various AFs is described as follows:

- The air coming out from lungs is obstructed in the vocal tract to produce a consonant sound unit. **Place of articulation** represents the point of contact between active and passive articulators in the vocal tract at which the obstruction occurs during the production of a consonant. The lower lip and tongue are the typical active articulators and the remaining articulators represent the passive articulators. There are eleven different places of articulation. For example, the bilabial sound units such as $/p/$ and $/b/$ are produced when the active lower lip comes in contact with passive upper lip.
- Different sound units are produced by obstructing air-stream in different ways with varying degrees of constriction. **Manner of articulation** represents the way in which the air-stream is obstructed and released from the vocal tract to produce a consonant. There are eight different manners of articulation. For example, in case of nasal sound units such as $/m/$ and $/n/$, air escapes from the vocal tract through nostrils.
- **Roundness** indicates whether the lips are rounded or not during the production of a vowel.
- **Frontness** indicates the horizontal position of the tongue during the articulation of a vowel relative to the front of the mouth.
- **Height** denotes the vertical position of the tongue during the production of a vowel relative to the aperture of the jaw [15].

There are mainly three ways to derive AFs: (i) acoustic-articulatory transformations using inverse mapping, (ii) direct physical measurements, and (iii) classification scores for pseudo-articulatory features [16, 17]. In the first approach, the articulatory movements are estimated from the speech acoustics through inverse mapping of speech. The process of inverse mapping refers to the inverse of the natural transformation from articulatory movements to speech acoustics and is referred to as the inverse problem or acoustic-to-articulatory inversion [18–21]. The second approach deals with capturing the motions of articulators through direct physical measurement techniques such as X-ray filming (cineradiography), magnetic resonance imaging [22–25], electromagnetic articulography [26–28], and electropalatography [29]. Use of these techniques require costly setup and involve the risk of health hazards such as exposure to radiations, etc. In the third approach, the statistical classifiers such as neural networks are used to derive AFs from spectral features. The spectral features from the acoustic signal are given as input and the classification scores are obtained at the output of the classifier [16, 30–32]. The classification scores thus obtained indicate pseudo-articulatory features [33, 34]. Among the three approaches, the third approach is popularly used and more feasible. Hence, this work uses third approach to derive the multilingual AFs, and then use them to improve the performance of Multi-PRSs."

1.3   Approaches for Multilingual Phone Recognition

There are mainly two approaches for multilingual phone recognition. First approach is LID-Mono approach where a front-end LID-switched to a monolingual phone recognizer trained individually on each of the languages present in multilingual dataset. In the second approach, a *common multilingual phone-set* that is derived based on the transcription of the multilingual dataset is used to develop multilingual phone recognition system.

The LID-Mono system developed using the first approach consists of two stages. *First stage* is a language identification stage used to detect the language of the input speech utterance. *Second stage* consists of monolingual phone recognizers of all the languages for which the multilingual phone recognition system is being developed. The multilingual phone recognition is carried out by detecting the language of the input speech utterance in the first stage, followed by phone recognition using the monolingual phone recognizer of the detected language in the second stage. This approach for multilingual phone recognition is more conventional and suffers due to the following issues: (i) Complex two-stage architecture, (ii) Failure of LID block leads to the failure of entire system, (iii) Developing monolingual phone recognizer is not feasible for all languages. This model is suitable for developing off-line or batch solution based systems as it suffers due to high latency period (response time).

In the second approach, Multi-PRS is developed using *common multilingual phone-set*. A universal acoustic model is trained by merging the training datasets of all the languages present in the multilingual dataset. The data shared by multiple languages is exploited to develop a common acoustic model. This approach enables the idea of *one model for all languages* leading to ease of management. It is less complex, more feasible to develop and more reliable compared to the more conventional LID-Mono approach. However, the development of such a *common multilingual phone-set* based Multi-PRS is faced with the specific difficulty of having to arrive at the appropriate common phone-set based on which such a phonetic decoding can be done on input speech from any of the languages of interest. Such a common phone-set has to have a coverage of all the phones occurring across the multiple languages while also ensuring that the individual language's phones are accurately mapped to the phones in the common multilingual phone-set. In this work, these two approaches for multilingual phone recognition in non-code-switched and code-switched scenarios are studied and compared using Indian languages.

1.4   Code-switched Phone Recognition using Multilingual Phone Recognition Systems

In code-switching, two or more languages are mixed and spoken as if they are one language [35, 36]. Code-switching (also referred as code-mixing or language-mixing or language alternation) involves switching between multiple languages either inter-sententially and intra-sententially [37]. Intra-sentential code-switching refers to the code-switching within a sentence, while the inter-sentential code-switching refers to the code-switching between the sentences. In code-switching, an utterance will have the lexical items and bounded morphemes from two or more languages [36]. Code-switching in children happens unconsciously due to their confusion in identifying between different languages, while code-switching by adults is a conscious use of blend of two languages [35, 37], for reasons such as, (i) availability of a better word or phrase in another language to express a particular idea, (ii) certain words or phrases are more readily available in the other language, and (iii) to show expertise in multiple languages. Bilingual code-switching, which refers to the fusion of two language sources, is more common compared to the mixing of more than two languages [38]. Hence, in this work, *bilingual code-switching* is studied in the context of Indian Languages using code-switched data of Kannada and Urdu languages.

Code-switching is a common practice across the world in multilingual societies such as India, where a speaker has spoken proficiency in more than one language, e.g. the mother tongue and one or more foreign languages. Typically geo-spatially neighbouring languages or those learnt for educational and professional reasons, e.g. an Indian language such as Kannada as mother tongue with proficiency in English (as a medium of instruction in education) or Urdu (being influenced by early Moghul rules and spoken by sections of the society spread across the country). The code-switching affects the co-articulation and context dependent acoustic modelling. Therefore, it is very challenging to develop ASR systems to perform code-switched speech recognition [10]. In this work, various techniques of multilingual phone recognition for decoding the code-switched speech data are examined.

1.5 Objective and Scope of the Work

The primary objective of this work is to study various aspects of Multi-PRSs such as development, performance improvement, and applications of Multi-PRSs in the context of Indian languages. Indian languages are characterized by a marked diversity even while enjoying significant shared underlying phonology and have not received much attention in the literature barring several monolingual continuous phone recognition approaches (for the purposes of developing phonetic engines in Indian languages [39–43]), and isolated word applications (for typical telephony information retrieval tasks in Indian languages [44–48]).

Exploiting this phonological commonality, this book focuses on several aspects of such a Multi-PRS, such as the use of IPA-derived common multilingual phone-set, use of predicted articulatory features, use of different learning frameworks, establishing a low PER for oracle AFs which sets the target performance reachable by proposed approaches, identifying the code-switching in Indian languages as a typical target application of such a multilingual phone recognizer, employing the proposed Multi-PRS for decoding code-switched speech (of *two* Indian languages), and comparing the conventional LID-switched monolingual approach with *common multilingual phone-set* based approach for multilingual phone recognition. These aspects that constitute the objectives, scope, and contributions of this work are highlighted further in the following:

- **IPA-derived multilingual phone-set:** A *common multilingual phone-set* is derived to phonetically decode each of the Indian languages being considered, specifically employing an IPA based transcription thereby facilitating a unified phonetic representation across all the languages. This helps to uniquely specify each phone in a language even while accounting for its possible shared identity in another language. This approach thereby differs from a few other earlier approaches for Indian languages, which have either examined the use of single-language phonetic representations or syllabic representations deemed invariant across languages.
- **Predicted articulatory features and feature fusion:** The AFs are predicted from spectral features via deep-learning based frameworks. The linguistic independence of such articulatory representation is ideally suited for this multilingual approach, wherein the above multilingual phone-set gets represented in the AF space, enhancing the discriminative capability of these units for any language and offering enhanced phone recognition performances over conventional spectral features. In addition to the early fusion of articulatory and spectral features using tandem approach, the use of phone posterior features shows a considerable promise in this multilingual setting, considering their complementary representational and discriminative ability.
- **Oracle AF performance:** In an important result, it is shown that the oracle AFs, that are derived from the ground truth IPA transcriptions, offer a performance of 10.4% PER, establishing the target performance reachable by the predicted AFs. If the AFs are predicted with high accuracy either from other spectral feature representations or via speech to articulatory inversion methods (an area currently still ongoing and holding much promise), the best performance achievable will be that of the oracle AFs.
- **Learning Frameworks:** In this work, the Multi-PRSs are primarily built using DNN-HMM frameworks, in addition to the more conventional GMM-HMM frameworks. The MTL framework is examined to improve the prediction accuracy of AF-predictors of various AF groups. We have explored various approaches for feature fusion (e.g. tandem, lattice rescoring), and different levels of combination of AF groups (e.g. consonant AFs, vowel AFs, all AFs) to generate the acoustic-models for Multi-PRSs.
- **Approaches for Multilingual Phone Recognition:** The LID-Mono and *common multilingual phone-set* based approaches for multilingual phone recognition are evaluated and compared. Specifically, the objective is to bring out the effectiveness and superiority of the proposed *common multilingual phone-set* based system over conventional LID-Mono approach. The LID-Mono approach heavily depends on the accuracy of the LID system and the LID errors cannot

be recovered. But, the *common multilingual phone-set* based system by virtue of not having to do a front-end LID switching and designed based on the common multilingual phone-set derived from several languages, is not constrained by the accuracy of the LID system, and hence performs effectively on non-code-switched and code-switched speech, offering low PERs than the LID-Mono system.

- **Code-switched Phone Recognition:** The code-switching scenario is an interesting and apt application domain to effectively put to use the multilingual phone recognition system. A code-switched continuous speech between two or more Indian languages (under consideration) can be seamlessly decoded (phonetically) using the common multilingual phone-set, thereby obviating the need to build and use monolingual phone recognition engines.

As summary, the scope of the work is to examine a suite of techniques and frameworks ideally suited for multilingual phone recognition each with its intrinsic multilingual appropriateness. In particular, for Indian languages, this work is a timely contribution, given the relevance of Indian languages for varied speech-based applications such as telephony information retrieval, smart-phone based voice-search applications, front-end to multilingual machine translation efforts, resolving the quintessentially multilingual problem of code-switching scenarios seamlessly without requiring individual monolingual engines.

1.6 Proposed Organization of the Book

- Chapter 1 provides brief introduction to multilingual speech recognition. The AFs, multilingual phone recognition approaches, and code-switched phone recognition are briefly introduced. The objective and scope of the present work are discussed. The chapter-wise organization is provided at the end of this chapter.
- Chapter 2 provides the literature survey. Prior work on multilingual speech recognition systems is provided. The use of AFs to improve performance of multilingual speech recognizers is briefly reviewed in this chapter. The existing works related to the use of multilingual systems for code-switched speech recognition are described. Various approaches used in the development of multilingual speech recognition systems are briefly described.
- Chapter 3 describes the development of Multi-PRS for four Indian languages. Detailed description on multilingual speech corpora and experimental setup is provided. The development and evaluation of Multi-PRS is presented. The results of Multi-PRSs and Mono-PRSs are compared and analysed with respect to Dravidian and Indo-Aryan language families. The performance improvement of Multi-PRS using tandem phone posterior features is also discussed.
- Chapter 4 discusses the proposed approaches to derive the multilingual AFs from spectral features. The development of AF-predictors for predicting the

AFs of five AF groups is discussed. The performance evaluation of various AF-predictors is described. The use of MTL framework to improve the prediction accuracy of AFs is discussed.

- Chapter 5 discusses the use of predicted multilingual AFs to improve the performance of Multi-PRS. The lattice rescoring method and tandem method of combination for combining the AFs from various AF groups is described. The *AF-based Multi-PRSs* are developed using predicted AFs and their performance is evaluated. The results are analysed and compared with their corresponding *oracle AF-based Multi-PRSs*.
- Chapter 6 describes the applications of multilingual phone recognition in code-switched and non-code-switched scenarios. It compares two approaches for multilingual phone recognition using code-switched and non-code-switched test sets. The development and evaluation of Multi-PRSs using LID-Mono and common multilingual phone-set based approaches is described. The analysis and comparison of the results is provided. The code-switched speech recognition using Multi-PRSs is studied using code-switched speech data of Kannada and Urdu languages.
- Chapter 7 summarizes the contributions of this book and provides future directions.

References

1. G.R. Tucker, A Global Perspective on Bilingualism and Bilingual Education, in *ERIC Digest* (Office of Educational Research and Improvement (ED), Washington, DC, 1999)
2. MHRD, Government of India, Language Education. [Online]. https://mhrd.gov.in/language-education [Accessed Mar. 08, 2020]
3. MHRD, Government of India, To know more about Indian Languages. http://mhrd.gov.in/sites/upload_files/mhrd/files/upload_document/languagebr.pdf [Accessed Mar. 08, 2020]
4. V. Golla, *California Indian Languages*. (University of California Press—Language Arts & Disciplines, California, 2011)
5. J. Cui et al., Multilingual representations for low resource speech recognition and keyword search, in *IEEE Workshop On Automatic Speech Recognition and Understanding* (2015), pp. 259–266. https://doi.org/10.1109/ASRU.2015.7404803
6. A. Waibel, H. Soltau, T. Schultz, T. Schaaf, F. Metze, Multilingual Speech Recognition, in *Verbmobil: Foundations of Speech-to-Speech Translation. Artificial Intelligence* (Springer, Berlin, 2000), pp. 33–45. https://doi.org/10.1007/978-3-662-04230-4_3
7. Y. Gao, B. Zhou, L. Gu, R. Sarikaya, H.-K. Kuo, A. I. Rosti, M. Afify, W. Zhu, IBM MASTOR: Multilingual Automatic Speech-to-Speech Translator, in *IEEE International Conference on Acoustics, Speech, and Signal Processing (ICASSP)* (2006), pp. 1205–1208. https://doi.org/10.1109/ICASSP.2006.1661498
8. N.T. Vu, D. Imseng, D. Povey, P. Motlicek, T. Schultz, H. Bourlard, Multilingual deep neural network based acoustic modeling for rapid language adaptation, in *IEEE International Conference on Acoustics, Speech, and Signal Processing (ICASSP), Florence* (2014), pp. 7639–7643. https://doi.org/10.1109/ICASSP.2014.6855086
9. A. Stolcke, M. Akbacak, L. Ferrer, S. Kajarekar, Improving Language Recognition with Multilingual Phone Recognition and Speaker Adaptation Transforms, in *Proceedings of the Odyssey 2010: The Speaker and Language Recognition Workshop* (2010), pp. 256–262

10. J. Weiner, N.T. Vu, D. Telaar, F. Metze, T. Schultz, D. Lyu, E. Chng, H. Li, Integration of language identification into a recognition system for spoken conversations containing code-switches, in *Proceedings of the 3rd Workshop on Spoken Language Technology for Under-resourced Languages(SLTU)* (2012)

11. N.T. Vu, D. Lyu, J. Weiner, D. Telaar, T. Schlippe, F. Blaicher, E. Chng, T. Schultz, L. Haizhou, A first speech recognition system for Mandarin-English code-switch conversational speech, in *IEEE International Conference on Acoustics, Speech, and Signal Processing (ICASSP)* (2012), pp. 4889–4892. https://doi.org/10.1109/ICASSP.2012.6289015

12. E. Yilmaz, H.V.D. Heuvel, D.V. Leeuwen, Investigating Bilingual Deep Neural Networks for Automatic Recognition of Code-switching Frisian Speech, in *Proceedings of the 5th Workshop on Spoken Language Technology for Under-resourced Languages(SLTU)* (2016), pp. 159–166. https://doi.org/10.1016/j.procs.2016.04.044

13. S. Stuker, F. Metze, T. Schultz, A. Waibel, Integrating Multilingual Articulatory Features Into Speech Recognition, in *INTERSPEECH* (2003), pp. 1033–1036

14. The International Phonetic Association, *Handbook of the International Phonetic Association* (Cambridge University, Cambridge, 2007). https://www.internationalphoneticassociation.org/ [Accessed Mar. 08, 2020]

15. Gerfen, *Phonetics Theory* (2011), pp. 251–257. http://www.unc.edu/~gerfen/Ling30Sp2002/phonetics.html [Accessed Apr. 08, 2016]

16. K. Kirchhoff, G.A. Fink, G. Sagerer, Combining acoustic and articulatory feature information for robust speech recognition. Speech Commun. **37**, 303–319 (2002). https://doi.org/10.1016/S0167-6393(01)00020-6

17. K.E. Manjunath, K.S. Rao, Improvement of phone recognition accuracy using articulatory features. Circuits Syst. Signal Process. (Springer) **37**(2), 704–728 (2017). https://doi.org/10.1007/s00034-017-0568-8

18. S. Dusan, L. Deng, Estimation of articulatory parameters from speech acoustics by Kalman filtering, in *Proceedings of CITO Researcher Retreat* (1998), pp. 47–48

19. H. Wakita, Direct estimation of the vocal tract shape by inverse filtering of acoustic speech waveforms, in *IEEE Transactions on Audio, Speech, and Language Processing* (1973), pp. 417–427. https://doi.org/10.1109/TAU.1973.1162506

20. S. Hahm, J. Wang, Parkinson's condition estimation using speech acoustic and inversely mapped articulatory data, in *INTERSPEECH* (2015), pp. 513–517

21. N. Dhananjaya, B. Yegnanarayana, S.V. Gangashetty, Acoustic-phonetic information from excitation source for refining manner hypotheses of a phone recognizer, in *IEEE International Conference on Acoustics, Speech and Signal Processing (ICASSP), Prague* (2011), pp. 5252–5255. https://doi.org/10.1109/ICASSP.2011.5947542

22. S. Narayanan, E. Bresch, P. Ghosh, L. Goldstein, A. Katsamanis, Y. Kim, A. Lammert, M. Proctor, V. Ramanarayanan, Y. Zhu, A multimodal real-time MRI articulatory corpus for speech research, in *INTERSPEECH* (2011), pp. 837–840

23. S. Narayanan, A. Toutios, V. Ramanarayanan, A.C. Lammert, J. Kim, S. Lee, K.S. Nayak, Y. Kim, Y. Zhu, L. Goldstein, D. Byrd, E. Bresch, P.K. Ghosh, A. Katsamanis, M.I. Proctor, Real-time magnetic resonance imaging and electromagnetic articulography database for speech production research (TC). J. Acoust. Soc. Am. **136**(3), 1307–1311 (2014). https://doi.org/10.1121/1.4890284

24. P.K. Ghosh, S. Narayanan, Automatic speech recognition using articulatory features from subject-independent acoustic-to-articulatory inversion. J. Acoust. Soc. Am. Express Lett. (2011), pp. 251–257. https://doi.org/10.1121/1.3634122

25. A. Afshan, P.K. Ghosh, Better acoustic normalization in subject independent acoustic-to-articulatory inversion: Benefit to recognition, in *International Conference on Acoustics, Speech and Signal Processing (ICASSP), Shanghai* (2016), pp. 5395–5399. https://doi.org/10.1121/1.3634122

26. S. Lee, S. Yildirim, A. Kazemzadeh, S. Narayanan, An articulatory study of emotional speech production, in *INTERSPEECH* (2005), pp. 497–500

27. The Centre for Speech Technology Research, The University of Edinburgh, MOCHA-TIMIT : MOCHA MultiCHannel Articulatory database: English. http://www.cstr.ed.ac.uk/research/projects/artic/mocha.html [Accessed Mar. 08, 2020]

28. D. Dash, M. Kim, K. Teplansky, J. Wang, Automatic Speech Recognition with Articulatory Information and a Unified Dictionary for Hindi, Marathi, Bengali, and Oriya, in *INTERSPEECH*, Hyderabad (2018). https://doi.org/10.21437/INTERSPEECH.2018-2122

29. C.S. Blackburn, Articulatory Methods for Speech Production and Recognition. PhD Thesis (Trinity College Cambridge and Cambridge University Engineering Department, Cambridge, 1996)

30. J. Frankel, M. Magimai-Doss, S. King, K. Livescu, O. Cetin, Articulatory Feature Classifiers Trained on 2000 hours of Telephone Speech, in *INTERSPEECH* (2007), pp. 2485–2488

31. O. Cetin, A. Kantor, S. King, C. Bartels, Magimai-Doss, J. Frankel, K. Livescu, An Articulatory Feature-Based Tandem Approach and Factored Observation Modeling, in *IEEE International Conference on Acoustics, Speech and Signal Processing (ICASSP-2007)*, Honolulu, HI (2007), pp. IV-645–IV-648. https://doi.org/10.1109/ICASSP.2007.366995

32. M. Rajamanohar, E. Fosler-Lussier, An evaluation of hierarchical articulatory feature detectors, in *IEEE Workshop on Automatic Speech Recognition and Understanding*, San Juan (2005), pp. 59–64. https://doi.org/10.1109/ASRU.2005.1566528

33. K.S. Rao, K.E. Manjunath, Speech recognition using articulatory and excitation source features, in *SpringerBriefs in Electrical and Computer Engineering book series* (Springer, Berlin, 2017). https://doi.org/10.1007/978-3-319-49220-9

34. K.E. Manjunath, Articulatory and excitation source features for phone recognition. *MS Thesis* (IIT Kharagpur, Kharagpur, WB, 2015)

35. S. Ford, *Language Mixing Among Bilingual Children*. http://www2.hawaii.edu/~sford/research/mixing.htm [Accessed Mar. 08, 2020]

36. J.F. Kroll, A.M.B.D. Groot, ed. *Handbook of Bilingualism: Psycholinguistic Approaches.* (Oxford University, Oxford, 2005)

37. L. Jorschick, A.E. Quick, D. Glasser, E. Lieven, M. Tomasello, German–English-speaking children's mixed NPs with correct agreement. Bilingualism: Language and Cognition **14**(2), 173–183 (2011). https://doi.org/10.1017/S1366728910000131

38. R.R. Heredia, J. Altarriba, Bilingual language mixing: why do bilinguals code-switch? Curr. Directions Psychol. Sci. **10**(5), 164–168 (2001). https://doi.org/10.1111/1467-8721.00140

39. B.D. Sarma, M. Sarma, M. Sarma, S.R.M. Prasanna, Development of Assamese phonetic engine: some issues, in *IEEE INDICON* (2013), pp. 1–6. https://doi.org/10.1109/INDCON.2013.6725966

40. K.E. Manjunath, K.S. Rao, D. Pati, Development of Phonetic engine for Indian languages: Bengali and Oriya, in *Proceedings of the Sixteenth IEEE International Oriental COCOSDA*, Gurgaon (2013), pp. 1–6. https://doi.org/10.1109/ICSDA.2013.6709900

41. K. Kumar, R.K. Aggarwal, Hindi Speech Recognition system using HTK. Int. J. Comput. Sci. Business Res. **2**(2), 12p. (2011)

42. M. Dua, R.K. Aggarwal, V. Kadyan, S. Dua, Punjabi automatic speech recognition using HTK. Int. J. Comput. Sci. Issues **9**(4), (2012)

43. M.V. Shridhara, B.K. Banahatti, L. Narthan, V. Karjigi, R. Kumaraswamy, Development of Kannada speech corpus for prosodically guided phonetic search engine, in *O-COCOSDA* (2013), pp. 1–6. https://doi.org/10.1109/ICSDA.2013.6709875

44. T.G. Yadava, H.S. Jayanna, A spoken query system for the agricultural commodity prices and weather information access in Kannada language. Int. J. Speech Technol. **20**(3), 635–644 (2017). https://doi.org/10.1007/s10772-017-9428-y

45. A. Dey, A. Deka, S. Imani, B. Deka, R. Sinha, S.R.M. Prasanna, P. Sarmah, K. Samudravijaya, S.R. Nirmala, AGROASSAM: a web based Assamese speech recognition application for retrieving agricultural commodity price and weather information, in *INTERSPEECH* (2018), pp. 3214–3215

46. S.G. Dontamsetti, P.K. Sahu, Speech based access of agricultural dealers information in Odia language. J. Appl. Theory Comput. Technol. **1**(1), 8–16 (2016). https://doi.org/10.22496/atct20161026106
47. H. Sailor, H.A. Patil, Neural networks-based automatic speech recognition for agricultural commodity in Gujarati language, in *Workshop on Spoken Language Technologies for Under-Resourced Languages* (2018), pp. 15–19. https://doi.org/10.21437/SLTU.2018-4
48. Mandi Project, *Speech-Based Automated Commodity Price and Weather Information Helpline for Twelve Indian States*. https://asrmandi.wixsite.com/asrmandi [Accessed Mar. 08, 2020]

Chapter 2
Literature Review

2.1 Introduction

In this chapter, the existing works on multilingual speech recognition are briefly discussed. Research in multilingual speech recognition is actively pursued from last two decades. Several important contributions related to multilingual speech recognition are reported in the last two decades. Some of the studies have attempted to improve the performance of multilingual speech recognition systems using AFs. But the number of studies on multilingual speech recognition in the context of Indian languages is very limited. Some of the notable works related to multilingual speech recognition systems involving the development, performance improvement using AFs, and applications in code-switched speech recognition are briefly discussed in this chapter. The organization of this chapter is as follows: Sect. 2.2 describes the prior work related to multilingual speech recognition. In Sect. 2.3, prior work related to the use of AFs to improve the performance of multilingual speech recognition systems is provided. Section 2.4 provides the prior work related to code-switched speech recognition using multilingual speech recognition systems. Section 2.5 summarizes this chapter.

2.2 Prior Work on Multilingual Speech Recognition

Research in multilingual speech recognition is very actively pursued in the last two decades. Several important contributions related to multilingual speech recognition are reported in the literature. Few notable works among them are described below. In 1998, Corredor-Ardoy et al. [1] developed a multilingual phone recognizer for spontaneous telephone speech using 4 languages—French, British English, German, and Castilian Spanish. HMMs and phonotactic bigram models are used to study the influence of the training material composition (size and linguistic content) on

© The Author(s), under exclusive license to Springer Nature Switzerland AG 2022
Manjunath K. E., *Multilingual Phone Recognition in Indian Languages*,
SpringerBriefs in Speech Technology, https://doi.org/10.1007/978-3-030-80741-2_2

the recognition performance. It is found that the training data consisting of only spontaneous speech data as linguistic content has shown the highest recognition performance compared to all other scenarios.

In 2000, Waibel et al. [2] developed a multilingual speech-to-speech translation system called *Verbmobil*. The Verbmobil consists of recognition engines of three languages—German, English, and Japanese—and a LID component that will switch between the recognizers. The combination of LID and recognition engines resulted in a flexible and user-friendly multilingual spoken dialog system.

In 2001, Schultz et al. [3–6] proposed methods for combining the multilingual acoustic models using a polyphone decision tree. Large Vocabulary Continuous Speech Recognition (LVCSR) systems for 15 languages having IPA transcription from GlobalPhone project are investigated. Similar IPA symbols from multiple languages are grouped to derive a common multilingual phone-set. Results are analysed for language dependent, independent, and language adaptive acoustic models.

In 2001, Uebler [7] presented various approaches for multilingual speech recognition. The different approaches namely, portation, cross-lingual, and simultaneous multilingual speech recognition are studied. The experiments are conducted to find the performance of cross-lingual speech recognition of an untrained language with a recognizer trained using other languages. The results of cross-lingual recognition for different baseline systems are compared and found that the number of shared acoustic units is very important for the performance.

In 2002, Ma et al. [8] studied the multilingual word recognition task using Mandarin and English languages. They proposed two approaches to deal with the performance degradation caused by the acoustic score biasing. The first approach uses a score normalization scheme by incorporating N-best scores from competing phone models into evaluating the normalized likelihood scores. The second approach uses a linear classifier to perform language identification followed by speech recognition.

In 2005, Kumar et al. [9] presented a unified approach for development of HMM based multilingual speech recognizer. Acoustically similar phones across multiple languages are grouped together using Bhattacharyya distance measure. The study considers two acoustically similar languages—Tamil and Hindi—along with an acoustically very different language, American English.

In 2005, Suryakant V G et al. [10] developed a syllable-based multilingual speech recognizer using three Indian languages—Tamil, Telugu, and Hindi.

In 2010, Burget et al. [11] presented an approach for multilingual speech recognition that uses *Subspace Gaussian Mixture Model*, where the state distributions are GMMs with a common structure, constrained to lie in a subspace of the total parameter space. The parameters that define this subspace can be shared across languages. It is shown that the proposed approach yields substantial improvements in the Word Error Rates (WER) of multilingual speech recognizers.

In 2012, Lin et al. [12] evaluated various architectures for multilingual speech recognition of real-time mobile applications. Experiments are conducted on a trilingual English–French–Mandarin recognition task for mobile applications. They explored various approaches for multilingual speech recognition such as (i) single large multilingual system based on data pooling from all the languages, (ii) using an explicit LID system to select the appropriate monolingual recognizer, and (iii) combining the confidence scores of monolingual and multilingual recognizers. It is shown that combination of results from several recognizers greatly outperforms all other solutions. The combined system is roughly 5% absolute better than an explicit LID approach, and 10% better than a single large multilingual system.

In 2013, Heigold et al. [13] trained multilingual acoustic models using DNNs and compared them with monolingual, cross-lingual systems. Eleven Romance languages with a total amount of 10 k hours of data is used for conducting experiments. It is found that multilingual systems outperform both cross-lingual and monolingual systems.

In 2013, Ramani et al. [14] proposed a uniform HMM framework for building language independent speech synthesis systems for six Indian languages, namely, Hindi, Marathi, Bengali, Tamil, Telugu, and Malayalam. The phonetic similarities among Indian languages are exploited to derive the language independent common phone-set and common question-set. The multilingual acoustic models are trained using the data of six Indian languages to hasten the segmentation process required for building HTS based speech synthesizers.

In 2014, Mohan et al. [15] developed a small vocabulary multilingual speech recognizer using two linguistically similar Indian languages—Hindi and Marathi. A subspace Gaussian mixture model is trained using the multilingual speech data collected for agricultural commodities task. A cross-corpus acoustic normalization technique is employed to handle speaker variations.

In 2014, Vu et al. [16] developed multilingual DNN-based acoustic modelling that can be applied to new languages. The effect of phone merging on multilingual DNN in the context of rapid language adaptation is investigated. Ten different languages from the Globalphone dataset are considered. It is found that the cross-lingual model transfer through multilingual DNN in combination with KL-HMM decoding yields the best performance.

In 2015, Muller et al. [17] described an approach for building LVCSR systems for under-resourced languages using the data from different languages. The language adaptive framework is developed using DNNs by feeding the language information directly to the network. The effectiveness of the proposed method is demonstrated through a series of experiments. The use of language dependent features called Language Feature Vectors (LFVs) has improved the performance of multilingual speech recognizers.

In 2015, Gonzalez-Domingueza et al. [18] presented an end-to-end multilingual speech recognition architecture developed and deployed at Google. The *Google Multilang Corpus* consisting of data from 8 languages with 1000 manually annotated utterances per language is used. The proposed system allows the users to speak in arbitrary combination of various languages (i.e. code-switching between multiple languages) and then performs ASR on it. The architecture consists of a front-end LID that identifies the language of the spoken utterance in real-time and then switches to the corresponding monolingual speech recognizer for decoding the spoken utterance. The LID is performed using the complementary information derived from the DNN-based LID classifier and the transcription confidences emitted by the monolingual speech recognizers. The system is evaluated using real-time data of 34 languages and found that the results are comparable with that of monolingual speech recognizers in terms of recognition accuracy and latency.

In 2017, Muller et al. [19] presented a method for training the *connectionist temporal classification* based multilingual speech recognition systems. The global phone-set is created by merging the pronunciation dictionaries from multiple languages. Experiments are conducted using Euronews corpus consisting of 10 languages with *Baidu's Deepspeech 2* architecture. It is shown that the performance of multilingual systems is improved by using the combination of LFVs and acoustic features.

In 2018, Toshniwal et al. [20] presented a single sequence-to-sequence (seq2seq) ASR model for multilingual speech recognition, which can recognize speech without requiring explicit language specification. The training data from 9 different Indian languages having very little overlap in their scripts is used. A union of language-specific grapheme sets is taken to train a grapheme-based sequence-to-sequence model jointly on data from all languages. The proposed model and its variants substantially outperform baseline monolingual sequence-to-sequence models for all languages and rarely choose the incorrect grapheme set in its output.

In 2018, Zhouhave et al. [21] investigated on the multilingual speech recognition of low-resource languages using single multilingual ASR Transformer. The dependency on the pronunciation lexicon is removed by choosing sub-words for multilingual modelling. The experiments are conducted using CALLHOME datasets.

In 2018, Cho et al. [22] presented a seq2seq approach for low-resource ASR. They used data from 10 BABEL languages to build a multilingual seq2seq model as a prior model, and then port them towards 4 other BABEL languages using transfer learning approach. The different architectures for improving the prior multilingual seq2seq model are explored. It is found that the transfer learning approach shows substantial gains over monolingual models across all 4 BABEL languages.

A summary of prior work on multilingual speech recognition is provided in Table 2.1.

Although there have been significant efforts in developing multilingual speech recognizers, the number of works exploring the development of multilingual speech recognizers in the context of Indian languages is very limited. While noting that no other multilingual effort has examined the use of IPA to derive a common multilingual phone-set labelling mechanism in the context of Indian languages, it is also to be noted that, the merging of the sound units from different languages in multilingual speech recognition works of Indian languages is limited to simplistic approaches such as syllable-based, Bhattacharyya distance based, and isolated word recognition based. In this study, the use of IPA based transcription to derive a *common multilingual phone-set* and a unifying framework that is easily generalizable to new languages is proposed. From the above works, it can be noted

Table 2.1 Summary of Prior Work on Multilingual Speech Recognition

Year	Dataset	Important contribution	Ref.
1998	French, British English, German, Spanish	The influence of the training material composition on recognition performance of multilingual phone recognizers is studied	[1]
2000	German, English, Japanese	The combination of the LID and recognition engines is used to develop Verbmobil multilingual speech-to-speech translation system	[2]
2001	GlobalPhone dataset	Combination of multilingual acoustic models and language adaptation techniques are studied	[3–6]
2001	Seven languages	Portation, cross-lingual, and simultaneous approaches for multilingual speech recognition are studied	[7]
2002	Mandarin, English	Score normalization and linear classification approaches to deal with the performance degradation due to acoustic score biasing are proposed	[8]
2005	Tamil, Hindi, TIMIT	Multilingual speech recognizer are developed by merging the phones based on Bhattacharyya distance measure	[9]
2005	Tamil, Telugu, Hindi	Development of syllable-based multilingual speech recognizer based on vowel onset points is studied	[10]
2010	English, German, Spanish	Multilingual speech recognizers based on subspace Gaussian mixture models are studied	[11]
2012	English, French, Mandarin	Various architectures for multilingual speech recognition of real-time mobile applications are studied	[12]
2013	11 Romance languages	DNN-based multilingual acoustic models are trained and compared with monolingual, cross-lingual systems	[13]
2013	Six Indian languages	Development of language independent speech synthesis systems using common phone-set and question-set	[14]
2014	Hindi, Marathi	A multilingual word recognizer is trained using subspace Gaussian mixture model	[15]
2014	Globalphone dataset	Rapid language adaptation and cross-lingual model transfer techniques are studied	[16]
2015	Euronews	LVCSR systems for under-resourced languages are built using the data from different languages	[17]
2015	Google Multilang Corpus	End-to-end multilingual speech recognition architecture developed and deployed at Google	[18]
2017	Euronews corpus	Multilingual speech recognition systems based on connectionist temporal classification are studied	[19]
2018	9 Indian languages	A single seq2seq ASR model for multilingual speech recognition is described	[20]
2018	CALLHOME dataset	Study of single multilingual ASR transformer for multilingual speech recognition of low-resource languages	[21]
2018	10 BABEL languages	Sequence-to-sequence approach for low-resource ASR is investigated	[22]

Table 2.2 Important takeaways from prior work on multilingual speech recognition

• The research in multilingual speech recognition is actively pursued in last two decades.
• There are very limited number of works related to multilingual speech recognition in the context of Indian languages.
• The use of IPA based transcription to derive common multilingual phone-set is not explored in the context of Indian languages.
• The number of works using DNNs to develop multilingual speech recognition systems in the context of Indian languages is limited.

that the use of DNNs is not much explored to develop multilingual systems in the context of Indian Languages. In this work, the use of DNNs to develop Multi-PRS is proposed.

Few notable takeaways from the prior work on multilingual speech recognition are provided in Table 2.2.

2.3   Prior Work on Multilingual Speech Recognition using Articulatory Features

Since the AFs are more universal [23–25] and less language dependent features compared to the conventional spectral features [26–28], they can be explored to improve the performance of multilingual speech recognizers. Although the AFs are widely explored to improve the performance of monolingual speech recognizers [29–34], only limited number of works exploring the AFs to improve the performance of multilingual speech recognizers are reported. The number of works exploring the use of AFs to improve the performance of multilingual speech recognizers in the context of Indian languages is very limited. Few notable works in this direction are as follows.

In 1997, Deng [23] proposed an integrated-multilingual speech recognizer framework mainly focusing on cross-language portability. The articulatory, acoustic, and auditory features are used for capturing the cross-language commonality. The AFs are derived from the dynamic properties of the vocal tract that are derived using the task-dynamic model. The tract variables such as upper and lower lips, jaw, tongue body, tongue tip, velum, glottal width, total lung force, supralaryngeal vocal tract volume, and vocal fold tension are considered for representing the AFs. The study aims to a build a universal speech recognizer that can be used across all the languages.

In 2003, Stuker et al. [35, 36] shown that the AFs derived from cross-lingual and multilingual AF detectors can significantly reduce the WER of HMM based speech recognizers. The AF detectors can compensate the inter-language variability. They considered five languages—Chinese Mandarin, German, Japanese, Spanish, and English—whose data was taken from GlobalPhone [37] and Wall Street Journal corpora. The transcription was derived using IPA symbols. It is found that the feature detectors that are trained using the multilingual AFs have higher classification accuracy compared to the feature detectors trained using the AFs of single language. The performance of speech recognizers based on multilingual AF detectors is superior compared to the speech recognizers based on monolingual AF detectors. The use of multilingual AFs has significantly reduced the WER of HMM based recognizer.

In 2007, Ore [38] developed AF detectors for detecting the AFs using GMMs and Multi-Layer Perceptrons (MLPs). English dataset from Wall street journal corpus and German, Spanish, and Japanese datasets from GlobalPhone corpora [37] are used. Multilingual AF detectors are developed using the data from all four languages. Four monolingual AF detectors, one for each language, are also developed. The outputs of the AF detectors are used as features for training HMM based phoneme recognizer. It is shown that the AFs output by the multilingual AF detectors perform better than that of the monolingual AF detectors. It is also found that the speech recognizers using AFs have higher performance compared to Mel-frequency cepstral coefficients.

In 2011, Rasipuram et al. [39] used MTL to improve the prediction accuracies of AF estimators. MLPs are used for training the AF estimators. It shown that the use of MTL derived AFs has shown significant improvement in the recognition accuracy of the phone recognizers compared to non-MTL derived AFs.

In 2016, Muller et al. [40] demonstrated the development of speech recognizers for low-resource languages using multilingual speech corpora. Fully connected feedforward neural networks are used for predicting the AFs. The datasets of four languages (English, French, German, and Turkish) from *Euronews* corpus are considered for training the multilingual speech recognizer using DNNs. Multilingual phone-set is derived by merging the IPA symbols from all the languages. English with only 10 h is considered as a low-resource language in addition to the large amount of data from French, German, and Turkish languages for system training. The use of AFs has significantly reduced the WER.

In 2017, Sahraeian et al. [28, 41, 42] worked on the adaptation of multilingual DNNs to a low-resource language. Articulatory-like features are extracted using a feature transformation technique called Intrinsic Spectral Analysis (ISA) manifold learning. Data of 9 different languages from GlobalPhone corpus [37] are considered in their studies. The language independent behaviour of spectral and ISA features is studied using 7 AFs—front vowel, back vowel, open vowel, plosive, labial, nasal, and fricative—that were extracted from the bottleneck layer of DNNs. It is found that the ISA features exhibit better overall language independent behaviour than spectral features. Several experiments were conducted under monolingual, cross-lingual, and multilingual settings to demonstrate the usefulness of ISA. It is shown that the ISA features have significant advantages compared to the traditional filter-bank features in multilingual and low-resource language scenarios.

In 2018, Dash et al. [43] explored the use of articulatory information to improve the ASR performance using four Indian languages namely—Hindi, Marathi, Bengali, and Oriya. Articulatory movements were recorded during speech production using an electromagnetic articulograph and trained together with acoustic features

Table 2.3 Summary of prior work on multilingual speech recognition using Articulatory Features

Year	Dataset	Important contribution	Ref.
1997	German, English, Japanese	An integrated-multilingual speech recognizer framework for cross-language portability is developed using articulatory, acoustic, and auditory features	[23]
2003	Mandarin, German, Japanese, Spanish, English	Cross-lingual and multilingual AFs are explored to reduce the WER of HMM based ASRs	[35, 36]
2007	GlobalPhone and Wall street journal corpus	The performance improvements of phoneme recognizers using multilingual and monolingual AFs is investigated	[38]
2011	TIMIT	The use of MTL to improve the prediction accuracies of AF estimators is studied	[39]
2016	Euronews corpus	The use of AFs in language adaptation techniques to improve the performance of multilingual speech recognizers is studied	[40]
2017	Globalphone dataset (9 languages)	The language independent behaviour of spectral and ISA features is studied using monolingual, cross-lingual, and multilingual scenarios	[28, 41, 42]
2018	Hindi, Marathi, Bengali, Oriya	The articulatory information derived from electromagnetic articulograph is used to improve the performance of multilingual, multimodal speech recognizer	[43]

to build automatic speech recognizers for these languages. ASR systems are trained using GMM-HMM, DNN-HMM, and *long short term memory recurrent neural network-HMM*. A multilingual, multimodal speech recognizer that was built based on a unified dictionary consisting of common and unique phonemes of all the four languages has shown significant reductions in the phoneme error rates.

A summary of the prior work on multilingual speech recognition using AFs is provided in Table 2.3. From above works, it can be noted that only very few work have explored the use of AFs to improve the performance of multilingual speech recognizers in the context of Indian languages. In most of the existing works, the AFs are derived using shallow neural networks [35, 38] and the use of DNNs to derive AFs is not widely explored in the literature (except in some studies like [32]). Hence, in this work, the use of AFs to improve the performance of multilingual speech recognizers in the context of Indian languages is examined. The DNNs are explored for predicting the multilingual AFs. This study is perhaps the first of its effort in the context of Indian languages in several fronts, such as the use of IPA based transcription to derive a common multilingual phone-set, the use of DNNs for training the Multi-PRSs and for predicting the AFs, the application of MTL for prediction of multilingual AFs. Few notable takeaways from prior work on multilingual speech recognition using AFs are provided in Table 2.4.

Table 2.4 Important takeaways from prior work on multilingual speech recognition using Articulatory Features

---

- The use of AFs to improve the performance of multilingual speech recognizers in the context of Indian languages is very limited.

- Most of the existing works use shallow neural networks to predict AFs, and the use of DNNs to predict the AFs is very limited.

- The AFs are widely explored in case of monolingual speech recognizers, and the use of AFs in case of multilingual speech recognizers is comparatively limited.

- In this work, the use of DNNs for predicting the multilingual AFs, the application of MTL for more accurate prediction of multilingual AFs, and the use of AFs to improve the performance of Multi-PRS in the context of Indian languages are proposed.

2.4   Prior Work on Code-Switched Speech Recognition using Multilingual Speech Recognition Systems

The multilingual speech recognition systems can be used for decoding the code-switched speech data. Few notable works on code-switched speech recognition using multilingual speech recognizers are as follows. In 2006, Lyu et al. [44] compared two approaches for recognizing the Taiwanese–Mandarin code-switching utterances. In the *multi-pass ASR approach*, three stages of processing —namely language boundary detection, language identification, and language dependent speech recognition—are integrated. They proposed a *one-pass ASR* approach which is based on common multilingual phone-set of IPA transcription. It is shown that the *one-pass ASR* approach can satisfactorily recognize the code-switched utterances without using language boundary detection and language identification blocks.

In 2012, Bhuvanagirir et al. [45] presented an approach for speech recognition of mixed language using English and Hindi languages. They constructed pronunciation lexicon and language model using a small mixed language text corpus of English and Hindi. Acoustic models of Hindi and English are used to perform phoneme recognition. Unlike the multi-pass approach for mixed language ASR, this approach does not require language boundary detection and language identification blocks.

In 2012, Weiner et al. [46] presented an approach for integrating LID into a multilingual ASR system for code-switched spoken conversations between Mandarin and English. Two approaches namely multi-stream and language look-ahead are introduced and compared. Both the approaches have shown improvements over the baseline results. The impact of LID accuracy on ASR performance is investigated. It is found that the higher LID accuracies will result in better ASR performance of code-switched speech.

In 2012, Vu et al. [47] presented a LVCSR system for conversational Mandarin–English code-switched speech recognition using SEAME corpus. Acoustic modelling is done by merging the phones based on the IPA and Bhattacharyya distance followed by discriminative training. The code-switched language models are built using the *statistical machine translation* approaches. An integrated multi-stream approach is used to provide the language information from LID to the decoding process.

In 2014, Schultz [48] presented a detailed review on various approaches for decoding the code-switched speech data. Various approaches including the LID-switched monolingual ASR and multilingual ASR along with their benefits and limitations are discussed in detail.

In 2016, Yilmaz et al. [49] presented a bilingual DNN-based ASR system that was trained for the recognition of code-switched speech data between Frisian and Dutch languages. Two different DNN architectures with language independent and language dependent targets are investigated using Frisian broadcast database. They demonstrated that bilingual DNN trained by merging the phones of both languages provides the best recognition performance.

In 2018, Kim et al. [50] proposed a language-universal end-to-end speech recognition system based on a universal character set and language-specific gating mechanism that can recognize the speech from any language used in training. The potential benefits of the proposed system are demonstrated for decoding the code-switched speech utterances of English and Spanish languages. For bilingual code-switched application, it is shown that the proposed end-to-end system performs better or comparable with the models trained on monolingual training data.

A summary of the prior work on code-switched speech recognition using multilingual speech recognition systems is provided in Table 2.5.

Few notable takeaways from prior work on code-switched speech recognition using multilingual speech recognition systems are provided in Table 2.6.

Table 2.5 Summary of prior work on code-switched speech recognition using multilingual speech recognition systems

Year	Dataset	Important contribution	Ref.
2006	Taiwanese, Mandarin	One-pass ASR and multi-pass ASR approaches for code-switched speech recognition are studied	[44]
2012	English, Hindi	The use of pronunciation lexicon and language models constructed from mixed language text corpus for code-switched speech recognition is studied	[45]
2012	Mandarin, English	The use of multilingual ASR with an integrated LID in it for code-switched speech recognition is studied	[46]
2012	SEAME corpus	LVCSR system for conversational code-switched speech recognition using an integrated multi-stream approach is studied	[47]
2014	Globalphone dataset	Various approaches for code-switched speech recognition including the LID-switched monolingual ASR and multilingual ASR are described	[48]
2016	Frisian, Dutch	Bilingual DNN-based ASR for code-switched speech recognition is studied	[49]
2018	English, Spanish	The language-universal end-to-end speech recognition system for code-switched speech recognition is studied	[50]

Table 2.6 Important takeaways from prior work on code-switched speech recognition using multilingual speech recognition systems

• There are very limited number of works exploring the code-switched speech recognition using multilingual speech recognition systems in the context of Indian languages.
• Mainly, two approaches of multilingual speech recognition—(i) LID-switched to monolingual speech recognition engines (i.e. multi-pass ASR) and (ii) Common multilingual phone-set based (i.e. one-pass ASR)—are explored for code-switched speech recognition.
• In this work, the decoding of code-switched speech data of Kannada and Urdu languages using Multi-PRS is studied.

2.5 Summary

In this chapter, overview of prior work on multilingual speech recognition, the existing work related to the use of AFs for multilingual speech recognition, and the related works on code-switched speech recognition using multilingual speech recognition systems are briefly described. There are very limited number of works related to multilingual speech recognition in the context of Indian languages. The use of AFs to improve the performance of multilingual speech recognizers in the context of Indian languages is very limited. There are very limited number of works exploring the use of multilingual speech recognition systems for decoding the code-switched speech data in the context of Indian languages. Hence, in this work, the development, performance improvement (using AFs), and applications (in code-switching) of Multi-PRSs are studied using Indian languages.

References

1. C. Corredor-Ardoy, L. Lamel, M. Adda-Decker, J.L. Gauvain, Multilingual phone recognition of spontaneous telephone speech, in *IEEE International Conference on Acoustics, Speech and Signal Processing (ICASSP)* (1998), pp. 413–416. https://doi.org/10.1109/ICASSP.1998. 674455
2. A. Waibel, H. Soltau, T. Schultz, T. Schaaf, F. Metze, Multilingual speech recognition, in *Verbmobil: Foundations of Speech-to-Speech Translation. Artificial Intelligence* (Springer, Berlin, 2000), pp. 33–45. https://doi.org/10.1007/978-3-662-04230-4_3
3. T. Schultz, A. Waibel, Language independent and language adaptive acoustic modeling for speech recognition. Speech Commun. **35**, 31–51 (2001). https://doi.org/10.1016/S0167-6393(00)00094-7
4. T. Schultz, A. Waibel, Language independent and language adaptive large vocabulary speech recognition, in *International Conference on Spoken Language Processing (ICSLP)* (1998), pp. 1819–1822
5. T. Schultz, A. Waibel, Multilingual and crosslingual speech recognition, in *Proceedings of the DARPA Workshop on Broadcast News Transcription and Understanding* (1998), pp. 259–262
6. T. Schultz, K. Kirchhoff, *Multilingual Speech Processing.* (Academic Press, New York, 2006). https://doi.org/10.1016/B978-0-12-088501-5.X5000-8
7. U. Uebler, Multilingual speech recognition in seven languages. Speech Commun. **35**(1–2), 53–69 (2001). https://doi.org/10.1016/S0167-6393(00)00095-9

8. B. Ma, C. Guan, H. Li, C. Lee, Multilingual speech recognition with language identification, in *Proceedings of the Seventh International Conference on Spoken Language Processing* (2002)

9. C.S. Kumar, V.P. Mohandas, L. Haizhou, Multilingual speech recognition: a unified approach, in *INTERSPEECH* (2005)

10. S.V. Gangashetty, C. Chandra Sekhar, B. Yegnanarayana, Spotting multilingual consonant-vowel units of speech using neural network models, in *Proceedings of the International Conference on Non-Linear Speech Processing (NOLISP)* (2005), pp. 303–317. https://doi.org/10.1007/11613107_27

11. L. Burget, P. Schwarz, M. Agarwal, P. Akyazi, K. Feng, A. Ghoshal, O. Glembek, N. Goel, M. Karafi, D. Povey, A. Rastrow, R.C. Rose, S. Thomas, Multilingual acoustic modeling for speech recognition based on subspace Gaussian mixture models, in *IEEE International Conference on Acoustics, Speech and Signal Processing (ICASSP)*, Dallas, TX (2010), pp. 4334–4337. https://doi.org/10.1109/ICASSP.2010.5495646

12. H. Lin, J.T. Huang, F. Beaufays, B. Strope, Y. Sung, Recognition of multilingual speech in mobile applications, in *IEEE International Conference on Acoustics, Speech, and Signal Processing (ICASSP)*, Kyoto (2012), pp. 4881–4884. https://doi.org/10.1109/ICASSP.2012.6289013

13. G. Heigold, V. Vanhoucke, A. Senior, P. Nguyen, M. Ranzato, M. Devin, J. Dean, Multilingual acoustic models using distributed deep neural networks, in *Proceedings of the IEEE International Conference on Acoustics, Speech and Signal Processing (ICASSP)*, Vancouver, BC (2013), pp. 8619–8623. https://doi.org/10.1109/ICASSP.2013.6639348

14. B. Ramani, S. Christina, G.A. Rachel, V.S. Solomi, M.K. Nandwana, A. Prakash, S.A. Shanmugam, R. Krishnan, S.P. Kishore, K. Samudravijaya, P. Vijayalakshmi, T. Nagarajan, H.A. Murthy, A common attribute based unified HTS framework for speech synthesis in Indian languages, in *Proceedings of the 8th ISCA Speech Synthesis Workshop* (2013), pp. 291–296

15. A. Mohan, R. Rose, S.H. Ghalehjegh, S. Umesh, Acoustic modelling for speech recognition in Indian languages in an agricultural commodities task domain. Speech Commun. **56**, 167–180 (2014). https://doi.org/10.1016/j.specom.2013.07.005

16. N.T. Vu, D. Imseng, D. Povey, P. Motlicek, T. Schultz, H. Bourlard, Multilingual deep neural network based acoustic modeling for rapid language adaptation, in *IEEE International Conference on Acoustics, Speech, and Signal Processing (ICASSP)*, Florence (2014), pp. 7639–7643. https://doi.org/10.1109/ICASSP.2014.6855086

17. M. Muller, A. Waibel, Using language adaptive deep neural networks for improved multilingual speech recognition, in *International Workshop on Spoken Language Translation (IWSLT)* (2015)

18. J.G. Dominguez, D. Eustis, I.L. Moreno, A. Senior, F. Beaufays, P.J. Moreno, A real-time end-to-end multilingual speech recognition architecture. IEEE J. Selected Topics Signal Process. **9**(4), 749–759 (2015). https://doi.org/10.1109/JSTSP.2014.2364559

19. M. Muller, S. Stuker, A. Waibel, Language Adaptive Multilingual CTC Speech Recognition, in *SPECOM 2017* (LNCS 10458), pp. 473–482 (2017). https://doi.org/10.1007/978-3-319-66429-3_47

20. S. Toshniwal, T.N. Sainath, R.J. Weiss, B. Li, P. Moreno, E. Weinstein, K. Rao, Multilingual speech recognition with a single end-to-end model, in *IEEE International Conference on Acoustics, Speech, and Signal Processing (ICASSP)*, pp. 4904–4908 (2018). https://doi.org/10.1109/ICASSP.2018.8461972

21. S. Zhou, S. Xu, B. Xu, Multilingual end-to-end speech recognition with a single transformer on low-resource languages, in *CoRR, vol. abs/1806.05059* (2018)

22. J. Cho, M.K. Baskar, R. Li, M. Wiesner, S.H. Mallidi, N. Yalta, M. Karafi, S. Watanabe, T. Hori, Multilingual sequence-to-sequence speech recognition: architecture, transfer learning, and language modeling, in *IEEE Spoken Language Technology Workshop (SLT)*, Athens, Greece (2018), pp. 521–527. https://doi.org/10.1109/SLT.2018.8639655

23. L. Deng, Integrated-multilingual speech recognition using universal phonological features in a functional speech production model, in *IEEE International Conference on Acoustics, Speech, and Signal Processing (ICASSP)*, Munich, vol.2 (1997), pp. 1007–1010. https://doi.org/10.1109/ICASSP.1997.596110

24. F. Metze, Articulatory features for conversational speech recognition, *Ph.D. Thesis* (Carnegie Mellon University, New York, 2005)

25. Y. Zhao, R. Zhao, X. Wang, Q. Ji, Multilingual articulatory features augmentation learning, in *Proceedings of the 23rd International Conference on Pattern Recognition (ICPR)*, Cancun (2016), pp. 2895–2899. https://doi.org/10.1109/ICPR.2016.7900076

26. K. Livescu, O. Cetin, M. Hasegawa-Johnson, S. King, C. Bartels, N. Borges, A. Kantor, P. Lal, L. Yung, A. Bezmaman, S. Dawson-Haggerty, B. Woods, J. Frankel, M. Magimai-Doss, and K. Saenko, Articulatory feature-based methods for acoustic and audio-visual speech recognition: summary from the 2006 JHU summer workshop, in *IEEE International Conference on Acoustics, Speech and Signal Processing (ICASSP)*, Honolulu, HI (2007), pp. IV-621–IV-624. https://doi.org/10.1109/ICASSP.2007.366989

27. A.W. Black, H.T. Bunnell, Y. Dou, P.K. Muthukumar, F. Metze, D. Perry, T. Polzehl, K. Prahallad, S. Steidl, C. Vaughn, Articulatory features for expressive speech synthesis, in *IEEE International Conference on Acoustics, Speech and Signal Processing (ICASSP)*, Kyoto (2012), pp. 4005–4008. https://doi.org/10.1109/ICASSP.2012.6288796

28. R. Sahraeian, D.V. Compernolle, Crosslingual and multilingual speech recognition based on the speech manifold. IEEE/ACM Trans. Audio, Speech, Language Process. **25**(12), 2301–2312 (2017). https://doi.org/10.1109/TASLP.2017.2751747

29. K. Kirchhoff, G.A. Fink, G. Sagerer, Combining acoustic and articulatory feature information for robust speech recognition. Speech Commun. **37**, 303–319 (2002). https://doi.org/10.1016/S0167-6393(01)00020-6

30. J. Frankel, M. Magimai-Doss, S. King, K. Livescu, O. Cetin, Articulatory feature classifiers trained on 2000 hours of telephone speech, in *INTERSPEECH* (2007), pp. 2485–2488

31. O. Cetin, A. Kantor, S. King, C. Bartels, M. Magimai-Doss, J. Frankel, K. Livescu, An articulatory feature-based tandem approach and factored observation modeling, in *IEEE International Conference on Acoustics, Speech and Signal Processing (ICASSP-2007)* , Honolulu, HI (2007), pp. IV-645–IV-648. https://doi.org/10.1109/ICASSP.2007.366995

32. V. Mitra, G. Sivaraman, H. Nam, C. Espy-Wilson, E. Saltzman, Articulatory features from deep neural networks and their role in speech recognition, in *IEEE International Conference on Acoustics, Speech and Signal Processing (ICASSP)*, Florence (2014), pp. 3017–3021. https://doi.org/10.1109/ICASSP.2014.6854154

33. R. Rasipuram, M. Magimai.-Doss, Articulatory feature based continuous speech recognition using probabilistic lexical modeling. Comput. Speech Language, vol. 36 (2016), pp. 233–259. https://doi.org/10.1016/j.csl.2015.04.003

34. K.E. Manjunath, K.S. Rao, Improvement of phone recognition accuracy using articulatory features. Circuits, Syst. Signal Processing, (Springer) **37**(2), 704–728 (2017). https://doi.org/10.1007/s00034-017-0568-8

35. S. Stuker, T. Schultz, F. Metze, A. Waibel, Multilingual articulatory features, in *IEEE International Conference on Acoustics, Speech and Signal Processing (ICASSP)*, vol. 1 (2003), pp. 144–147. https://doi.org/10.1109/ICASSP.2003.1198737

36. S. Stuker, F. Metze, T. Schultz, A. Waibel, Integrating multilingual articulatory features into speech recognition, in *INTERSPEECH* (2003), pp. 1033–1036

37. T. Schultz, Globalphone: a multilingual speech and text database developed at Karlsruhe university, in *ICSLP-2002*, Denver, CO, USA (2002), pp. 345–348

38. B.M. Ore, Multilingual articulatory features for speech recognition, Master's thesis (Wright State University, New York, 2007)

39. R. Rasipuram, M. Magimai-Doss, Improving articulatory feature and phoneme recognition using multitask learning, in *Artificial Neural Networks and Machine Learning (ICANN)*, vol. 6791 (2011), pp. 299–306. https://doi.org/10.1007/978-3-642-21735-7_37

40. M. Muller, S. Stuker, A. Waibel, Towards improving low-resource speech recognition using articulatory and language features, in *International Workshop on Spoken Language Translation (IWSLT)* (2016), pp. 1–7.

41. R. Sahraeian, *Acoustic Modeling of Under-resourced Languages*, Ph.D. Thesis (Katholieke Universiteit Leuven (KU Leuven), Leuven, 2017)

42. R. Sahraeian, D.V. Compernolle, F.D. Wet, On using intrinsic spectral analysis for low-resource languages, in *Proceedings of the 4th International Workshop on Spoken Language Technologies for Under-resourced Languages (SLTU)* (Saint-Petersburg, Russia, 2014), pp. 61–65.

43. D. Dash, M. Kim, K. Teplansky, J. Wang, Automatic speech recognition with articulatory information and a unified dictionary for Hindi, Marathi, Bengali, and Oriya, in *INTERSPEECH*, Hyderabad (2018). https://doi.org/10.21437/INTERSPEECH.2018-2122

44. D. Lyu, R. Lyu, Y. Chiang, C. Hsu, Speech recognition on code-switching among the Chinese dialects, in *IEEE International Conference on Acoustics, Speech, and Signal Processing (ICASSP)*, Toulouse (2006), pp. I–I. https://doi.org/10.1109/ICASSP.2006.1660218

45. K. Bhuvanagirir, S.K. Kopparapu, Mixed language speech recognition without explicit identification of language. Am. J. Signal Process. **2**(5), 92–97 (2012). https://doi.org/10.5923/j.ajsp.20120205.02

46. J. Weiner, N.T. Vu, D. Telaar, F. Metze, T. Schultz, D. Lyu, E. Chng, H. Li, Integration of language identification into a recognition system for spoken conversations containing code-switches, in *Proceedings of the 3rd Workshop on Spoken Language Technology for Under-resourced Languages(SLTU)* (2012)

47. N.T. Vu, D. Lyu, J. Weiner, D. Telaar, T. Schlippe, F. Blaicher, E. Chng, T. Schultz, L. Haizhou, A first speech recognition system for Mandarin-English code-switch conversational speech, in *IEEE International Conference on Acoustics, Speech, and Signal Processing (ICASSP)* (2012), pp. 4889–4892. https://doi.org/10.1109/ICASSP.2012.6289015

48. T. Schultz, Multilingual automatic speech recognition for code-switching speech, in *The 9th International Symposium on Chinese Spoken Language Processing*, Presentation (2014)

49. E. Yilmaz, H.V.D. Heuvel, D.V. Leeuwen, Investigating Bilingual deep neural networks for automatic recognition of code-switching Frisian speech, in *Proceedings of the 5th Workshop on Spoken Language Technology for Under-resourced Languages(SLTU)* (2016), pp. 159–166. https://doi.org/10.1016/j.procs.2016.04.044

50. S. Kim, M. L. Seltzer, Towards language-universal end-to-end speech recognition, in *IEEE International Conference on Acoustics, Speech and Signal Processing (ICASSP)* (2018), pp. 4914–4918. https://doi.org/10.1109/ICASSP.2018.8462201

Chapter 3
Development and Analysis of Multilingual Phone Recognition System

3.1 Introduction

In the previous chapter, prior work related to multilingual speech recognition is discussed. In this chapter, the development of Multi-PRS using four Indian languages—Kannada, Telugu, Bengali, and Odia—is described. IPA based transcription is used for grouping the acoustically similar phonetic units from multiple languages. Multi-PRSs in the context of Indian languages are studied using two language families namely—Dravidian and Indo-Aryan. Both HMMs and DNNs are explored for training the PRS under both context dependent and context independent setups. The performance of Multi-PRSs is analysed and compared with that of the Mono-PRS. The advantages of Multi-PRSs over monolingual PRSs are discussed. The tandem Multi-PRSs are developed using phone posteriors as tandem features to improve the performance of the baseline Multi-PRSs. This chapter is organized as follows: Sect. 3.2 describes the experimental setup used in this work. Section 3.3 discusses about the development of phone recognition systems. Section 3.4 provides the details of performance evaluation of PRSs. In Sect. 3.5, the results of various monolingual and multilingual PRSs are compared and analysed. Section 3.6 summarizes the contents of this chapter.

3.2 Experimental Setup

In this study, MFCCs are used as features for building the PRSs. Both HMMs and DNNs are explored for training the systems. The following subsections provide the detailed description of the experimental setup used in this study.

Manjunath K. E., *Multilingual Phone Recognition in Indian Languages*,
SpringerBriefs in Speech Technology, https://doi.org/10.1007/978-3-030-80741-2_3

3.2.1 Multilingual Speech Corpora

The speech corpora of four Indian languages namely—Kannada, Telugu, Bengali, and Odia—is considered. The speech corpora was developed as a part of consortium project titled *Prosodically guided phonetic engine for searching speech databases in Indian languages* supported by DIT, Govt. of India [1]. Detailed description of speech corpora is provided in [2–11].

Speech corpora contains the speech wave files along with their IPA transcription [2, 12, 13]. The audio files were recorded using a sampling rate of 16 KHz and a precision of 16 bits per sample. The phonetically and prosodically rich transcription is derived using IPA chart for all the wave files. IPA chart provides a standardized representation of the sounds, independent of their language. IPA chart is designed based on the principle of strict one-to-one correspondence between the sounds and the symbols. An IPA symbol will have same production characteristics irrespective of the language in which it is produced. The sound units from any language can be represented using IPAs. This makes the IPA transcription to be more accurate in merging the acoustically similar phonetic units from multiple languages compared to the ASCII transcription. IPA chart consists of 70 (59 pulmonic and 11 non-pulmonic) consonants, 28 vowels, 31 diacritics, 19 additional signs, and 12 other symbols. The suprasegmental qualities such as stress, duration, intonation, and tone are indicated using *additional signs*. The vowels and consonants are modified using diacritics. Although the IPA chart contains around 160 symbols, a particular language can be represented by using very less number of symbols [13].

IPA transcription process is a very expensive, highly labour-intensive task, and requires high degree of skill and efforts. It is a skilled job and takes large number of human hours. Since it was observed that the quality of IPA transcription produced by the native speakers was better compared to non-native speakers, native speakers of a particular language were used for transcribing the speech of that specific language. The transcribers were trained initially and then allowed to do actual transcription of the speech. In order to derive IPA transcription, the transcribers used to carefully listen to the speech at phrase level and then transcribe the phrase into sequence of IPAs. In case of difficulties, the same portion of speech was listened multiple times and transcription was derived. Once the first level of transcription is complete, it was verified by another transcriber and corrected if there are any mistakes [2].

Although the speech corpora contains data in read, extempore, and conversation modes of speech [2, 14]; in this work, only read speech corpora is considered. The source of read speech includes the news broadcasts from television, radio channels, and the speech data collected from the speakers reading the textbooks or story-books in a noiseless closed room environment. The speech data is organized in the form of sentences. The duration of each wave file varies from 3 s to 11 s. The non-overlapping speakers are used for training and testing. A split of 80:20 is used for train and test data, respectively. 10% of training data is held out from the training and used as a development set.

Table 3.1 Statistics of multilingual speech corpora

Language	# Speakers		Duration (in hours)			
	Male	Female	Train	Dev	Test	Total
Kannada	7	9	2.80	0.33	0.76	3.89
Telugu	9	10	4.05	0.47	1.07	5.59
Bengali	20	30	3.42	0.40	0.99	4.81
Odia	14	16	3.58	0.36	0.97	4.91

Table 3.1 shows various statistics of multilingual speech corpora used in this study. The number of speakers and the duration of speech data are shown separately for each language. First column lists the names of the languages. Next two columns provide the count of male and female speakers, while the fourth to seventh columns tabulate the duration of different datasets in terms of number of hours.

3.2.2 Extraction of Mel-frequency Cepstral Coefficients

MFCCs are extracted from the given input signal using the standard procedure pescribed in [15]. A frame duration of 25 ms with a frame shift of 10 ms is used. 13-dimensional MFCCs along with their first and second order derivatives are computed for each frame. Cepstral mean and variance normalization is applied per-speaker basis on MFCCs followed by the transformation using linear discriminant analysis.

3.2.3 Training HMMs and DNNs

"Initially flat-start initialization is used to build Context Independent (CI) GMM-HMMs (referred as HMMs throughout). The alignments generated by the CI HMMs are used to initialize the training of Context Dependent (CD) HMMs. This is further followed by training the CD DNN-HMMs (referred as DNNs throughout) using the alignments obtained from the CD HMMs. CI DNNs are also trained using the alignments generated by the CI HMMs. The CI models are based on monophones, while the CD models are based on triphones. The mapping from phonetic context and the HMM-state index, to an emission probability density function is captured through acoustic-phonetic decision tree [16]. Number of Gaussians, number of transition states, and number of transition ids depend on the number of phones and context being modelled. For example, baseline Multi-PRS using KN, TE, BN, OD languages with 44 phones had 974 Gaussians, 132 transition states, and 264 transition ids with CI HMMs, while the CD HMMs for the same system had 15,039 Gaussians, 2078 transition states, and 4156 transition ids.

DNNs with tanh non-linearity at hidden layers and softmax activation at the output layer are used. DNNs are trained using greedy layer-by-layer supervised training. Initial learning rate was chosen to be 0.015 and was decreased exponentially for the first 15 epochs. A constant learning rate of 0.002 was used in the last 5 epochs. Once all the hidden layers are added to the network, shrinking is performed after every 3 iterations, so as to separately scale the parameters of each layer. Mixing up was carried out in the halfway between the completion of addition of all the hidden layers and the end of training. Stability of the training is maintained through preconditioned affine components. Once the final iteration of training completes, the models from last 10 iterations are combined into a single model. Each input to DNNs uses a temporal context of 9 frames (4 frames on either side) [17–20]. The number of hidden layers used in DNNs are tuned by adjusting the width of the hidden layers. It is found that the DNNs with 5 hidden layers are suitable for training Multi-PRSs. DNNs with dimensions of 432 input, 300 hidden, and 19,860 output layers are used for training the baseline quadrilingual Multi-PRS. The total number of parameters of the DNNs range between 1.9 million to 2.0 million based on the dimension of the features used. The size of the input layer depends on the dimension of features used for training the DNNs.

Bi-phone (phoneme bi-grams) language model is used for decoding input speech signal. The language model weighting factor and acoustic scaling factor used for decoding the lattice are optimally determined using the development set to minimize the PERs. DNNs training used in this study is similar to the one presented in [21]. Similar procedure is used for training DNNs in [22–25]. All the experiments are conducted using the open-source speech recognition toolkit—Kaldi [26].

3.3 Development of Phone Recognition Systems

Four Indian languages—Kannada, Telugu, Bengali, and Odia—are considered to study the multilingual phone recognition in the context of Indian languages. Four Mono-PRSs, one for each of the considered language are developed. Three Multi-PRSs—two bilingual PRSs and one quadrilingual PRS—are developed. The following subsections provide the detailed description of the development of Mono and Multi-PRSs.

3.3.1 Development of Monolingual Phone Recognition Systems

This section describes the procedure used for developing the Mono-PRSs. Mono-PRSs are developed using the data of a single language. Four Mono-PRSs, one for each of the four languages (KN, TE, BN, and OD) considered in this study, are developed. Since sufficient amount of data is not available to train a separate model for all the IPA symbols present in the transcription, only a limited set of phones

Table 3.2 Number of Consonants, Vowels, and Silence present in the phone-set considered by different Mono-PRSs

Mono-PRS	Count of Different Phonetic units			
	Consonants	Vowels	Silence	Total
Kannada	25	10	1	36
Telugu	23	11	1	35
Bengali	27	6	1	34
Odia	30	5	1	36

are considered for training. The limited set of phones is determined based on the frequency of occurrences of phones in the IPA transcription. The IPA symbols with high frequency of occurrence are considered as the candidates for limited set of phones. Each IPA symbol is mapped to the acoustically closest phone available in the considered set of phones. It is found that except : (i.e. long duration) and ^{h} (i.e. aspiration) IPA symbols, none of the diacritics or additional signs had sufficient number of occurrences to model them. Hence, all the diacritics and additional signs except : and ^{h} are ignored, and the phones with these ignored diacritics and additional signs are treated as simple base phones [27, 28]. Table 3.2 shows the phone-level statistics used in development of different PRSs. First column indicates the name of Mono-PRSs. Second to fifth columns show the number of consonants, vowels, silence, and the total number of phones, respectively for each Mono-PRS.

From Table 3.2, it is observed that the Dravidian languages have higher number of vowels compared to Indo-Aryan languages and vice versa for consonants. The higher number of vowels in Dravidian languages is due to the presence of long vowels in addition to short vowels, while most of the long vowels are missing in Indo-Aryan languages. This is because the long vowels in Dravidian languages had sufficient data to build a separate model for each of them. The *eleven* vowels in Telugu are due to the presence of /ɛ/ vowel in addition to all short and long vowels. The /ɛ/ is open-mid, front, unrounded vowel [13]. All the aspirated stop-consonants were present in Indo-Aryan languages, while some of the aspirated stop-consonants were missing in Dravidian languages. This resulted in higher number of consonants in Indo-Aryan PRSs. The following aspirated consonants were not modelled in KN and TE PRSs {/Th/, /chh/, /gh/, /ph/, /jh/, /Dh/}. This is because, the native speakers of Bengali and Odia are more accurate in pronouncing aspirated consonants than those of Kannada and Telugu, and vice versa for long vowels. The difference could be clearly perceived by making the native speakers of all four languages to utter the words such as /ratha/ or /rath/ (meaning chariot) and /khaDga/ or /khargo/ (meaning sword). Both HMMs and DNNs are explored for training Mono-PRSs using MFCCs under CI and CD settings. HMMs and DNNs are trained using the procedure mentioned in Sect. 3.2.3.

3.3.2 Development of Multilingual Phone Recognition Systems

Multi-PRSs are developed using the data from multiple languages. Three Multi-PRSs, with two being bilingual PRSs and one being a quadrilingual PRS, are developed. The *KN-TE-BN-OD* quadrilingual PRS is developed by merging the data from four languages—Kannada, Telugu, Bengali, and Odia. The bilingual PRSs are developed by grouping the linguistically similar languages. Dravidian languages (i.e. Kannada and Telugu) are grouped together to develop *KN-TE* bilingual PRS, while the Indo-Aryan languages (i.e. Bengali and Odia) are grouped together to develop *BN-OD* bilingual PRS. Since majority of the Indian languages come under Dravidian and Indo-Aryan language families, separate bilingual PRS for each of these groups is developed and their characteristics are analysed.

The *common multilingual phone-set* for each of the Multi-PRS is derived by merging the acoustically similar IPA symbols across all of its languages together and selecting the phonetic units which have sufficient number of occurrences to train a separate model for each of them. The IPAs which do not have sufficient number of occurrences will be mapped to the closest linguistically similar phonetic units present in the common multilingual phone-set. For example, the /gh/ phone in Kannada and Telugu languages had inadequate number of occurrences to a train a separate model for it, and hence it was mapped to its linguistically closest phone /g/. Similarly, /ee/ phone in Bengali and Odia languages is mapped to its linguistically closest phone /e/. Both HMMs and DNNs are explored for developing Multi-PRSs using MFCCs under CD and CI settings. HMMs and DNNs are trained using the procedure mentioned in Sect. 3.2.3. Table 3.3 shows the phone-level statistics used in development of various Multi-PRSs. First two columns list the type and name of Multi-PRSs. The type represents the type of Multi-PRS either bilingual or quadrilingual. Last four columns show the number of consonants, vowels, silence, and total number of phones, respectively.

Figure 3.1 illustrates the training of a Multi-PRS. Acoustically similar IPAs from all the languages are merged together. The common multilingual phone-set is derived from these merged IPAs. Feature extraction block will extract the features from the input speech signals. The extracted features along with their phonetic labels are used for training HMMs and DNNs. A multilingual decoder is used for decoding the phones present in the test utterance.

Table 3.3 Number of Consonants, Vowels, and Silence present in the common multilingual phone-set considered by different Multi-PRSs

Multi-PRS		Count of Different Phonetic units			
Type	Name	Consonants	Vowels	Silence	Total
Bilingual	KN-TE	28	11	1	40
	BN-OD	31	7	1	39
Quadrilingual	KN-TE-BN-OD	32	11	1	44

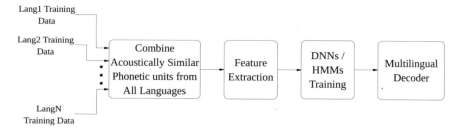

Fig. 3.1 Illustration of Training of a Multilingual Phone Recognition System

3.3.3   Development of Tandem Multilingual Phone Recognition Systems

This section describes the development of tandem Multi-PRSs to improve the performance of the baseline Multi-PRSs. Tandem approach is a data-driven signal processing method for extracting the features and then using them for acoustic modelling of the speech [29, 30]. The difference between the baseline and tandem Multi-PRSs is that the baseline Multi-PRSs are developed using MFCCs, whereas the tandem Multi-PRSs are developed using the combination of MFCCs and tandem features. Tandem features are derived from the spectral features by training a classifier. Tandem features provide supplementary information to perform more accurate classification. Different varieties of features such as articulatory and discriminative features can be captured in the form of tandem features. In this work, the discriminative features containing the information to discriminate between various phonetic units are used as tandem features. Discriminative features are the posterior probabilities produced by a discriminative classifier using the spectral features. It has been widely reported in the literature that the combination of spectral and tandem features will significantly improve the performance of speech recognizers [31–33]. In this work, the use of tandem features is explored under multilingual framework.

Figure 3.2 shows the block diagram of the development of tandem Multi-PRS. DNNs are used as a discriminative classifiers. A DNN for phone recognition is trained in the first stage using the MFCCs extracted from the training data. The posterior probabilities obtained at the output of first stage are called Phone Posteriors (PPs). The PPs are combined with the MFCCs to train a DNN-based tandem Multi-PRS in the second stage [34]. Multilingual decoder is used to decode the test utterance into a sequence of phones. The posterior probabilities of different phones in each frame is given by: $P(i) = p(q_t = i|x_t)$, where q_t is a phone at time $t, i = 1, 2 \ldots N$, and x_t is the spectral features at time t such that

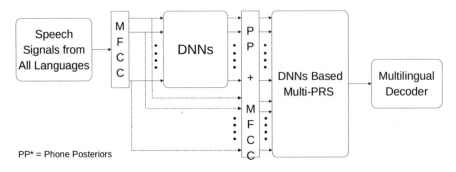

Fig. 3.2 Block diagram of Tandem Multilingual Phone Recognition System

$$\sum_{i=1}^{N} P(i) = 1,$$

$$where\ N = Total\ number\ of\ phone\ classes. \tag{3.1}$$

3.4   Performance Evaluation of Phone Recognition Systems

Sclite tool [35] is used for scoring and evaluating the output of PRSs. PER is determined by comparing the hypothesized text output by the speech recognizer with the reference (or correct) transcriptions using Dynamic Programming (DP). DP is used to get the optimal string alignment by minimizing the distance function between the reference and hypothesized phone pairs. DP string alignment algorithm performs a global minimization of a Levenshtein distance function which weights the cost of correct phones, insertions (I), deletions (D), and substitutions (S) as 0, 3, 3, and 4, respectively. PER in percentage is computed using Eq. (3.2).

$$Phone\ Error\ Rate = \frac{S + D + I}{N} \times 100\%, \tag{3.2}$$

where N is the total number of phones present in the reference transcriptions.

From each language, 20% of the non-overlapping data (see Table 3.1) is considered as the *test set* of that language. The performance of a Mono-PRS is evaluated using the test set of the language used for training it. The *test sets* of all the languages that are used for training a Multi-PRS are used as the *test set* for evaluating that specific Multi-PRS. For example, the performance of *KN-TE* bilingual PRS is found using the test sets of KN and TE languages. Likewise, the test sets of KN, TE, BN, and OD languages are used as the test set for evaluating the performance of *KN-TE-BN-OD* quadrilingual PRS. The following subsections provide the detailed description and analysis of the results obtained by various PRSs.

Table 3.4 Phone error rates of monolingual phone recognition systems

Mono-PRSs	Test sets	PERs (%)			
		CI		CD	
		HMM	DNN	HMM	DNN
Kannada	KN	43.5	39.5	38.5	37.1
Telugu	TE	42.1	35.5	35.0	30.7
Bengali	BN	49.0	41.6	43.4	37.6
Odia	OD	33.6	29.5	28.0	26.5
Average	–	42.1	36.5	36.2	33.0

3.4.1   Performance Evaluation of Monolingual Phone Recognition Systems

Table 3.4 shows the PERs of Mono-PRSs that are developed using MFCCs. First column shows the language used for building the Mono-PRSs. Second column shows the test set used for evaluating the performance. Third and fourth columns provide the PERs of the CI systems using HMMs and DNNs, respectively. Similarly, the last two columns tabulate the results of CD systems. The last row shows the average PERs of all the four Mono-PRSs. As one moves from left to right PERs decrease in all the rows. This indicates that the CD models have lower PERs than CI models. In all the cases, DNNs have lower PERs compared to the PERs of HMMs. The CD DNNs have shown least PERs in all Mono-PRSs.

3.4.2   Performance Evaluation of Multilingual Phone Recognition Systems

Table 3.5 shows the PERs of Multi-PRSs that are developed using MFCCs. First column shows the name of the Multi-PRS. Second column shows the test set used for evaluating the performance. Third and fourth columns provide the PERs of the CI systems using HMMs and DNNs, respectively. Similarly, the last two columns tabulate the results of CD systems. The PERs of Multi-PRSs decrease as one moves from CI models to CD models, and from HMMs to DNNs. Multi-PRSs are trained using the data pooled from multiple languages. This makes the data used in the development of Multi-PRS to have relatively higher number examples compared to the Mono-PRSs, which are trained using the data from a single language. Since DNNs require higher amount of data for training, the larger number of data samples available in case of Multi-PRSs are exploited by the DNNs to train more accurate models compared to the DNNs of Mono-PRSs that have comparatively smaller amount of data samples. This clearly shows the advantage of using DNNs for developing Multi-PRSs. The *KN-TE* bilingual PRS has lower PER (33.1%) compared to the average PER of KN and TE Mono-PRSs (33.9%), whereas the PER of *BN-OD* bilingual PRS (32.0%) and the average PER of BN and OD Mono-PRSs (32.05%) are almost the same.

Table 3.5 Phone error rates of multilingual phone recognition systems

Multi-PRSs	Test sets	PERs (%)			
		CI		CD	
		HMM	DNN	HMM	DNN
KN-TE	KN-TE	45.1	38.1	37.3	33.1
BN-OD	BN-OD	43.9	37.3	37.2	32.0
KN-TE-BN-OD	KN-TE-BN-OD	49.4	39.8	39.0	35.1

Further, a detailed analysis of the results obtained using *KN-TE-BN-OD* Multi-PRS is carried out. For simplicity and better readability, *KN-TE-BN-OD* Multi-PRS is referred as *Quadrilingual PRS (Quadri-PRS)*. The average of the PERs of all the four Mono-PRSs shown in the last row of Table 3.4 is termed as *Average PRS*. In case of CD models, the difference in the PERs of DNNs and HMMs is 3.9% and 3.2%, respectively, for *Quadri-PRS* and *Average PRS*. Similarly, CI models have a difference of 9.6% and 5.6%, respectively, for *Quadri-PRS* and *Average PRS*. This shows that the reductions in the PERs of *Average PRS* using DNNs are not as high as that of *Quadri-PRS*. This indicates that the improvement obtained using DNNs in *Quadri-PRS* is higher compared to the improvement obtained using DNNs in *Average PRS*. This is because the *Quadri-PRS* simultaneously uses the data shared from four languages. The *Quadri-PRS* using CD DNNs outperform Kannada and Bengali Mono-PRSs. It is found that vowels are better modelled by *KN* and *TE* Mono-PRSs, while the consonants are more accurately modelled in *BN* and *OD* Mono-PRSs. The *Quadri-PRS* takes the simultaneous mutual advantage of all the four languages and results in more accurate models for both consonants and vowels.

3.4.3 Performance Evaluation of Tandem Multilingual Phone Recognition Systems

Table 3.6 shows the PERs of tandem Multi-PRSs that are developed using the combination of MFCCs and PPs. First column shows the name of the tandem Multi-PRS. Second column shows the test set used for evaluating the performance. Third and fourth columns provide the PERs of the CI systems using HMMs and DNNs, respectively. Similarly, the last two columns tabulate the results of CD systems. It is observed that the tandem Multi-PRSs outperform the corresponding baseline Multi-PRSs (see Table 3.5) in all cases. This clearly shows that the use of tandem features has led to the improvement in the performance of tandem Multi-PRSs. For example, the use of tandem features in *KN-TE-BN-OD* system has led to 1% absolute PER reduction from 35.1% (baseline) to 34.1% (tandem).

Table 3.6 PERs of tandem multilingual phone recognition systems developed using the combination of MFCCs and phone posteriors

Tandem Multi-PRSs	Test sets	PERs (%)			
		CI		CD	
		HMM	DNN	HMM	DNN
KN-TE	KN-TE	35.6	34.3	35.7	31.9
BN-OD	BN-OD	33.3	32.5	33.1	30.6
KN-TE-BN-OD	KN-TE-BN-OD	36.6	35.4	36.5	34.1

3.5 Discussion of Results

This section discusses the detailed analysis of the results obtained. The results of Multi-PRSs are compared with that of Mono-PRSs. The cross-lingual analysis is carried out to gain further insights into the results obtained.

3.5.1 Analysis and Comparison of the Results

From Tables 3.2 and 3.3, it is observed that Multi-PRSs have higher number of phones compared to that of Mono-PRSs. This indicates that the Multi-PRSs could recognize more number of phones compared to that of Mono-PRSs. The Indian languages are analysed at two broad groups namely—Dravidian languages and Indo-Aryan languages. The *KN, TE,* and *KN-TE PRSs* are termed as Dravidian PRSs, whereas the *BN, OD,* and *BN-OD PRSs* are termed as Indo-Aryan PRSs. From Sect. 3.3.2, it is observed that the long vowels are more prominent in Dravidian languages, while the aspirated consonants are more prominent in Indo-Aryan languages. It is found that all the long vowels could not be recognized using Indo-Aryan PRSs, while the Dravidian PRSs cannot recognize all the aspirated stop-consonants. This makes the long vowels to be more accurately modelled in Dravidian PRSs and the aspirated consonants to be better modelled by Indo-Aryan PRSs. The mutual advantage of both Dravidian and Indo-Aryan PRSs is exploited in *Quadri-PRS*. Using the proposed *Quadri-PRS*, it is possible to recognize all the long vowels and aspirated consonants in both Dravidian and Indo-Aryan languages. This demonstrates the advantage of Multi-PRS over the Mono-PRS. The adequate occurrences of {/Th/, /chh/, /gh/, /ph/} phones are not found in the Dravidian PRSs, but are found in *Quadri-PRS*. Likewise, the adequate occurrences of {/ee/, /ii/, /uu/, /oo/, /v/} phones are not found in Indo-Aryan PRSs, but are found in *Quadri-PRS*. Figure 3.3 provides the Venn diagram representing the distribution of certain phones across Dravidian, Indo-Aryan, and Quadrilingual PRSs. The region marked by dark black lines belongs to *Quadri-PRS* and represents the union of Dravidian and Indo-Aryan PRSs.

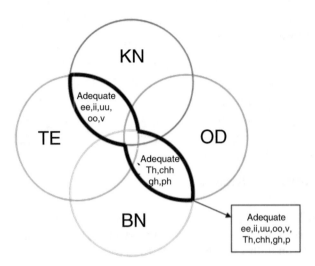

Fig. 3.3 Venn diagram representing the distribution of certain phones across Dravidian, Indo-Aryan, and Quadrilingual PRSs

It is observed that the phones /f/ and /ṣ/ were not present either in *KN* or *TE* Mono-PRS, but they were found in *KN-TE* Multi-PRS. This is because sufficient amount of data was not available in *KN* or *TE* Mono-PRSs to train separate models for /f/ and /ṣ/. However, merging the IPA symbols from Kannada and Telugu languages together has led to sufficient number of occurrences to build separate models for /f/ and /ṣ/. Similarly, the vowel /ɛ/ is not found either in *BN* or *OD* Mono-PRS, whereas it is present in *BN-OD* bilingual Multi-PRS. In Odia, /sh/ had only 52 occurrences, which were not adequate to build a separate model for /sh/. Hence, /sh/ was mapped to the acoustically closest phone /s/. This made the *OD* Mono-PRS incapable of identifying /sh/ phones. Since the Bengali had around 9800 /sh/ occurrences, after merging the /sh/ phonetic units from both Odia and Bengali, there were adequate number of /sh/ phones to develop a separate /sh/ model in *BN-OD* bilingual PRS. Now, it is possible to recognize the /sh/ units present in a Odia speech utterance using *BN-OD* bilingual PRS. This is one of unique advantage offered by the Multi-PRSs. Figure 3.4 provides the Venn diagram representing the distribution of certain phones across Monolingual and Bilingual PRSs.

Further, the causes for misclassifications are analysed at language and phone levels. If the number of examples contributed by a language towards training a phone are higher, then the misclassifications due to that specific phone from that particular language will be lower, and vice versa. For example, the contribution of Odia language to the training data of */sh/* model of BN-OD Multi-PRS is only 0.5%. This indicates that the */sh/* phones occur very rarely in Odia. Although none of the */sh/* occurrences was present in the test data of Odia, but many fricatives (more prominently */s/*) are decoded as */sh/*. This increased the overall misclassification rate of */sh/*. Similarly, the contribution of Odia language to the training data of /ŋ/

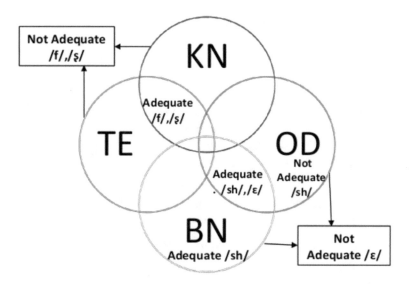

Fig. 3.4 Venn diagram representing the distribution of certain phones across Monolingual and Bilingual PRSs

in BN-OD Multi-PRS is only 2.3%, but there are many misclassifications of Odia test utterances to /ɲ/. This is one of the disadvantages of decoding using Multi-PRSs (i.e. *BN-OD* Multi-PRS) over Mono-PRSs (i.e. OD PRS).

Figure 3.5 compares the PERs of baseline and tandem Multi-PRSs. X-axis denotes the type of Multi-PRSs, and Y-axis indicates the PERs of Multi-PRSs. The baseline systems are represented using solid lines, whereas the tandem systems are shown in dashed lines. The legend indicates the colours of the lines corresponding to each Multi-PRS. All the lines corresponding to tandem systems are present below the lines corresponding to baseline systems indicating that the tandem Multi-PRSs have lower PERs compared to the baseline systems. All the lines have least value at CD DNNs indicating that all the Multi-PRSs have least PERs when trained using CD DNNs.

3.5.2   Cross-Lingual Analysis

Cross-lingual systems are trained using the data of one language and used for performing the recognition of another language. The cross-lingual approach is very useful in development of speech recognizers for low-resource languages. An attempt to carry out cross-lingual analysis on all the baseline and tandem Multi-PRSs is made and is as follows. The *four* test sets that are used in the evaluation of *four* Mono-PRSs and *three* test sets that are used in the evaluation of *three* Multi-PRSs, totalling *seven* different test sets are used to perform cross-lingual analysis. Since

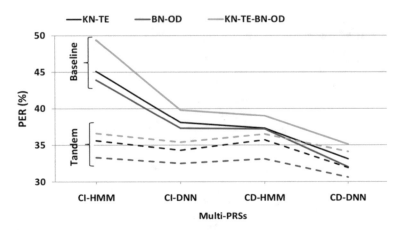

Fig. 3.5 Comparison of PERs of Baseline and Tandem Multilingual Phone Recognition Systems

the CD DNNs have shown highest performance (least PERs) in all the experiments carried out till now, only CD DNNs based PRSs (i.e. Mono-PRSs or Multi-PRSs) are used in all of the upcoming experiments.

Table 3.7 shows the results of Multi-PRSs for *seven* different test sets. The results are shown separately for baseline and tandem Multi-PRSs. If a test set belongs to a language, whose data is used in the development of a Multi-PRS then let us call it as *native-test set* otherwise it is referred as a *foreign-test set*. It is found that the test sets of the languages that are used in the development of a Multi-PRS have better performance (lower PER) compared to the test sets of other languages and vice versa. This means the *native-test sets* have lower PER compared to that of *foreign-test sets*. For example, the *Kannada* test set (shown in row 2, 10) performs better with KN-TE and KN-TE-BN-OD Multi-PRSs as *KN* is a *native-test set* for these systems, and performs poorly when used as a *foreign-test set* against BN-OD bilingual PRS. Similarly, when both *BN* and *OD* are used as test sets (see row 7, 15), they perform poorly with KN-TE bilingual PRS, and perform better with BN-OD and KN-TE-BN-OD Multi-PRSs. The KN-TE-BN-OD Multi-PRS shown in column 4 performs consistently well across all the test sets demonstrating the effectiveness of the *common multilingual phone-set* based on which it is developed. There is a clear reduction in the PERs of all the tandem Multi-PRSs compared to their corresponding baseline Multi-PRSs indicating the effectiveness of tandem features.

The results are analysed by computing the % change (absolute) and % change (relative) in the recognition accuracies of Multi-PRSs with respect to the monolingual PRSs. The recognition accuracy of a PRS can be derived from the PER using Eq. (3.3).

$$Recognition\ Accuracy(\%) = 100 - PER(\%). \tag{3.3}$$

Table 3.7 Cross-lingual analysis of baseline and tandem multilingual phone recognition systems

Test set	KN-TE	BN-OD	KN-TE-BN-OD
Baseline Multi-PRS			
KN	35.8	54.2	38.3
TE	31.6	48.7	34.4
BN	65.8	38.3	40.5
OD	44.2	26.2	28.7
KN+TE	33.1	50.9	34.3
BN+OD	54.6	32.0	35.8
KN+TE+BN+OD	43.9	41.4	35.1
Tandem Multi-PRS			
KN	34.1	53.3	37.9
TE	30.7	48.1	33.5
BN	65.3	36.9	39.3
OD	43.4	25.0	27.5
KN+TE	31.9	50.2	35.0
BN+OD	54.0	30.6	33.1
KN+TE+BN+OD	43.0	40.4	34.1

The % change (absolute) in the recognition accuracy of Multi-PRS with respect to the Mono-PRS is computed by taking the difference in the recognition accuracy of *Mono-PRS* and *Multi-PRS* for a given test set. % change (absolute) is given by Eq. (3.4).

$$\% \ Change \ (Absolute) \ = \ R.A \ (PRS_{Mono}) \ - \ R.A \ (PRS_{Multi}), \qquad (3.4)$$

where R.A stands for recognition accuracy.

The % change (relative) in the recognition accuracy of a *Multi-PRS* with respect to the *Mono-PRS* is defined as the ratio of the difference in the recognition accuracies of *Mono-PRS* and *Multi-PRS* to the recognition accuracy of *Mono-PRS* for a given test set. In a more simplified way, % change (relative) is the ratio of % change (absolute) to the recognition accuracy of *Mono-PRS* for a given test set. The % change (relative) is given by Eq. (3.5).

$$\% \ Change \ (Relative) \ = \ \frac{(R.A \ (PRS_{Mono}) - R.A \ (PRS_{Multi}))}{R.A \ (PRS_{Mono})} \times 100\%,$$

$$(3.5)$$

where R.A stands for recognition accuracy.

A negative value of % change (absolute) and % change (relative) indicates that Multi-PRSs outperform Mono-PRSs, whereas a positive value indicates vice versa of it.

Table 3.8 shows the % change (absolute) and % change (relative) in the recognition accuracies of Multi-PRSs with respect to the Mono-PRSs. First column indicates the type of Multi-PRSs (baseline or tandem). Second column shows the name of Multi-PRS. Last four columns list four different Mono-PRSs and test sets.

Table 3.8 The % Change (Absolute) and % Change (Relative) in the recognition accuracies of Multi-PRSs with respect to Mono-PRSs

Type	Multi-PRSs	Monolingual PRSs and Test sets			
		KN	TE	BN	OD
% Change (Absolute)					
Baseline (MFCCs)	KN-TE	−1.30	0.90	28.20	17.70
	BN-OD	17.10	18.00	0.70	−0.30
	KT-TE-BN-OD	1.20	3.70	2.90	2.20
Tandem (MFCCs + PPs)	KN-TE	−3.00	**0.00**	27.70	16.90
	BN-OD	16.20	17.40	−0.70	−1.50
	KN-TE-BN-OD	0.80	2.80	1.70	1.00
% Change (Relative)					
Baseline (MFCCs)	KN-TE	−2.07	1.30	45.19	24.08
	BN-OD	27.19	25.97	1.12	−0.41
	KT-TE-BN-OD	1.91	5.34	4.65	2.99
Tandem (MFCCs + PPs)	KN-TE	−4.77	**0.00**	44.39	22.99
	BN-OD	25.76	25.11	−1.12	−2.04
	KN-TE-BN-OD	1.27	4.04	2.72	1.36

For better analysis and comparison, the results are tabulated separately for baseline and tandem Multi-PRSs. The % change (absolute) and % change (relative) of both baseline and tandem Multi-PRSs are computed with respect to baseline Mono-PRSs, and are shown separately.

Let us consider KN-TE Multi-PRS against KN Mono-PRS and test set shown in row 3 and column 3. The KN is used as test set to determine the PERs of KN Mono-PRS and KN-TE Multi-PRS. The % change (absolute) is determined using Eq. (3.4), and the value obtained will be populated in row 3 and column 3. Similarly, the % change (relative) is determined using Eq. (3.5).

Higher the value (in Table 3.8) more the difference in recognition accuracies and poorer the performance of Multi-PRSs. Lower the value (in Table 3.8) lesser the difference in recognition accuracies and better the performance of Multi-PRSs. A value of *"0.00"* (see highlighted values) indicates that the recognition accuracies of both monolingual and multilingual PRSs are same. It can be observed that all the tandem Multi-PRSs have smaller values than their corresponding baseline Multi-PRSs. Generally, if the same dataset is used for training of both monolingual and Multi-PRSs, then such entries will have lower *% change (relative)* values and vice versa.

Figures 3.6 and 3.7 illustrate the variation of % change (absolute) and % change (relative) in the recognition accuracies of Multi-PRSs with respect to Mono-PRSs. In both the figures, X-axis denotes Mono-PRSs, and the Y-axis indicates % change. The baseline systems are represented using solid lines, whereas the tandem systems are shown in dashed lines. Both Figs. 3.6 and 3.7 follow exactly similar pattern. This is because, the % change (relative) is derived from % change (absolute). However, the scales of % change on Y-axis are different in graphs of Figs. 3.6 and 3.7.

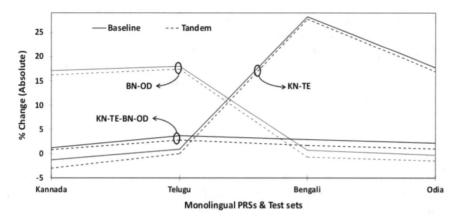

Fig. 3.6 Illustration of Variation of % Change (Absolute) in the Recognition Accuracies of Multi-PRSs with respect to Mono-PRSs

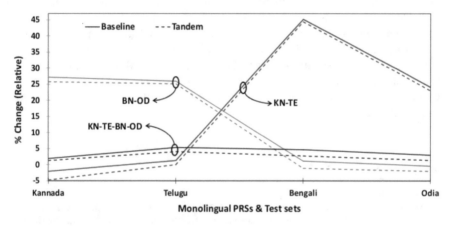

Fig. 3.7 Illustration of Variation of % Change (Relative) in the Recognition Accuracies of Multi-PRSs with respect to Mono-PRSs

If both the Multi-PRS and Mono-PRS that are being compared use the dataset from the same language, then the dataset is referred as *native-dataset*, otherwise it is called as *foreign-dataset*. It can be clearly observed that the lines related to *native-datasets* are present at the bottom compared to the lines related to *foreign-datasets* that are present at the top. For *KN-TE* Multi-PRS, the Kannada and Telugu are the *native-datasets*, while the Bengali and Odia are the *foreign-datasets*. It is found that in all the cases of *KN-TE* Multi-PRS, the lines corresponding to Kannada and Telugu (i.e. *native-datasets*) are present in the bottom, and the lines related to Bengali and Odia (i.e. *foreign-datasets*) are present in the top. Similar behaviour is observed in case of *BN-OD* Multi-PRS. A clear variation can be observed in both *KN-TE* and *BN-OD* Multi-PRS where the values vary between low to high depending on the *native* or *foreign* datasets. However, not much variation is observed in the lines representing the *KN-TE-BN-OD* Quadri-PRSs. This is because, the datasets of all the four Mono-PRSs are used in the development of Quadri-PRS (i.e. all are *native-datasets*). The lines representing the *KN-TE-BN-OD* Quadri-PRSs are very close to "0" (see *KN-TE-BN-OD* rows of Table 3.8). This indicates that the performance of the Multi-PRSs is close to that of Mono-PRSs.

3.6   Summary

The development of monolingual and multilingual PRSs is demonstrated using four Indian languages—Kannada, Telugu, Bengali, and Odia. It is found that the long vowels are better modelled by Dravidian PRSs, while the Indo-Aryan PRSs model the aspirated consonants more accurately. The Quadri-PRS that is developed using both Dravidian and Indi-Aryan languages takes the mutual advantage of both Dravidian and Indo-Aryan languages and better models both vowels and consonants. DNNs have superior performance compared to HMMs. CD systems outperform CI systems in all the cases. The CD DNNs have shown least PERs (i.e. highest performance) in all the cases. The performance of Multi-PRSs is improved using tandem features. The *KN-TE, BN-OD*, and *KN-TE-BN-OD* tandem Multi-PRSs have shown an absolute PER reduction of 1.2%, 1.4%, and 1.0%, respectively, compared to their baseline systems.

References

1. *Development of Prosodically Guided Phonetic Engine for Searching Speech Databases in Indian Languages*. http://speech.iiit.ac.in/svldownloads/pro_po_en_report/ [Accessed Mar. 08, 2020]
2. S.B.S. Kumar, K.S. Rao, D. Pati, Phonetic and prosodically rich transcribed speech corpus in Indian languages: Bengali and Odia, in *Proceedings of the Sixteenth IEEE International Oriental COCOSDA*, Gurgaon (2013), pp. 1–5. https://doi.org/10.1109/ICSDA.2013.6709901
3. M.V. Shridhara, B.K. Banahatti, L. Narthan, V. Karjigi, R. Kumaraswamy, Development of Kannada speech corpus for prosodically guided phonetic search engine, in *O-COCOSDA* (2013), pp. 1–6. https://doi.org/10.1109/ICSDA.2013.6709875
4. M.C. Madhavi, S. Sharma, H.A. Patil, Development of language resources for speech application in Gujarati and Marathi, in *IEEE International Conference on Asian Language Processing (IALP)*, vol. 1 (2014), pp. 115–118. https://doi.org/10.1109/IALP.2014.6973517
5. B.D. Sarma, M. Sarma, M. Sarma, S.R.M. Prasanna, Development of Assamese phonetic engine: some issues, in *IEEE INDICON* (2013), pp. 1–6. https://doi.org/10.1109/INDCON.2013.6725966
6. R.R. Kiran, S.B.S. Kumar, K.E. Manjunath, B. Satapathy, A. Chaturvedi, D. Pati, K.S. Rao, Automatic phonetic and prosodic transcription for Indian languages: Bengali and Odia, in *Proceedings of the 10th International Conference on Natural Language Processing (ICON)* (2013)

7. K.S. Rao, K.E. Manjunath, Speech recognition using articulatory and excitation source features, in *SpringerBriefs in Electrical and Computer Engineering Book Series* (Springer, Berlin, 2017). https://doi.org/10.1007/978-3-319-49220-9

8. K.E. Manjunath, Articulatory and excitation source features for phone recognition, MS Thesis (IIT Kharagpur, Kharagpur, WB, 2015)

9. K.E. Manjunath, K.S. Rao, Articulatory and excitation source features for speech recognition in read, extempore and conversation modes. Int. J. Speech Technol. (Springer) **19**(1), 121–134 (2016). https://doi.org/10.1007/s10772-015-9329-x

10. K.E. Manjunath, K.S. Rao, M. Gurunath Reddy, Improvement of phone recognition accuracy using source and system features, in *International Conference on Signal Processing and Communication Engineering Systems (SPACES)* (2015), pp. 501–505. https://doi.org/10.1109/SPACES.2015.7058205

11. K.E. Manjunath, S. Sunil Kumar, D. Pati, B. Satapathy, K.S. Rao, Development of consonant-vowel recognition systems for Indian languages: Bengali and Odia, in *IEEE India Conference on Emerging Trends and Innovation in Technology (INDICON)* (2013), pp. 1–6. https://doi.org/10.1109/INDCON.2013.6726109

12. K.E. Manjunath, K.S. Rao, D. Pati, Development of Phonetic Engine for Indian languages: Bengali and Oriya, in *Proceedings of the Sixteenth IEEE International Oriental COCOSDA*, Gurgaon (2013), pp. 1–6. https://doi.org/10.1109/ICSDA.2013.6709900

13. The International Phonetic Association, *Handbook of the International Phonetic Association* (Cambridge University, Cambridge, 2007). https://www.internationalphoneticassociation.org/ [Accessed Mar. 08, 2020]

14. K.E. Manjunath, K.S. Rao, Automatic phonetic transcription for read, extempore and conversation speech for an Indian language: Bengali, in *IEEE National Conference on Communications (NCC)* (2014), pp. 1–6. https://doi.org/10.1109/NCC.2014.6811347

15. L. Rabiner, B. Juang, B. Yegnanarayana, *Fundamentals of Speech Recognition*. (Pearson Education, New York, 2008)

16. K.T. Riedhammer, T. Bocklet, A. Ghoshal, D. Povey, Revisiting semi-continuous hidden Markov models, in *ICASSP* (2012), pp. 4721–4724. https://doi.org/10.1109/ICASSP.2012.6288973

17. P. Swietojanski, J. Li, J-T. Huang, Investigation of max out networks for speech recognition, in *IEEE International Conference on Acoustics Speech and Signal Processing (ICASSP)* (2014), pp. 7649–7653. https://doi.org/10.1109/ICASSP.2014.6855088

18. M. Cai, Y. Shi, J. Liu, Deep maxout neural networks for speech recognition, in *ASRU* (2013), pp. 291–296. https://doi.org/10.1109/ASRU.2013.6707745

19. D. Garcia-Romero, X. Zhang, A. McCree, D. Povey, Improving speaker recognition performance in the domain adaptation challenge using deep neural networks, in *IEEE Spoken Language Technology Workshop (SLT)* (2014), pp. 378–383. https://doi.org/10.1109/SLT.2014.7078604

20. J. Trmal, G. Chen, D. Povey, S. Khudanpur, P. Ghahremani, X. Zhang, V. Manohar, C. Liu, A. Jansen, D. Klakow, D. Yarowsky, F. Metze, A keyword search system using open source software, in *IEEE Spoken Language Technology Workshop (SLT)* (2014), pp. 530–535. https://doi.org/10.1109/SLT.2014.7078630

21. X. Zhang, J. Trmal, D. Povey, S. Khudanpur, Improving deep neural network acoustic models using generalized maxout networks, in *ICASSP* (2014), pp. 215–219. https://doi.org/10.1109/ICASSP.2014.6853589

22. J. Tao, S. Ghaffarzadegan, L. Chen, K. Zechner, Exploring deep learning architectures for automatically grading non-native spontaneous speech, in *IEEE International Conference on Acoustics Speech and Signal Processing (ICASSP)* (2016), pp. 6140–6144. https://doi.org/10.1109/ICASSP.2016.7472857

23. C. Liu, A. Jansen, S. Khudanpur, Context-dependent point process models for keyword search and detection-based ASR, in *IEEE International Conference on Acoustics Speech and Signal Processing (ICASSP)* (2016), pp. 6025–6029. https://doi.org/10.1109/ICASSP.2016.7472834

24. Z. Lv, M. Cai, C. Lu, J. Kang, L. Hui, W. Zhang, J. Liu, Improved system fusion for keyword search, in *IEEE Workshop on Automatic Speech Recognition and Understanding (ASRU)* (2015), pp. 230–236. https://doi.org/10.1109/ASRU.2015.7404799

25. A. Schwarz, C. Huemmer, R. Maas, W. Kellermann, Spatial diffuseness features for DNN-based speech recognition in noisy and reverberant environments, in *IEEE International Conference on Acoustics Speech and Signal Processing (ICASSP)* (2015), pp. 4380–4384. https://doi.org/10.1109/ICASSP.2015.7178798

26. D. Povey et al., The Kaldi speech recognition toolkit, in *IEEE Workshop on ASRU* (2011). http://kaldi-asr.org/ [Accessed Mar. 08, 2020]

27. K.E. Manjunath, K.S. Rao, D.B. Jayagopi, Development of multilingual phone recognition system for Indian languages, in *IEEE International Conference on Signal Processing, Informatics, Communication and Energy Systems (SPICES)*, Kollam (2017), pp. 1–6. https://doi.org/10.1109/SPICES.2017.8091271

28. K.E. Manjunath, D.B. Jayagopi, K.S. Rao, V. Ramasubramanian, Development and analysis of multilingual phone recognition systems using Indian languages, in *International Journal of Speech Technology, (Springer)* (2019), pp. 1–12. https://doi.org/10.1007/s10772--018-09589-z

29. O. Cetin, A. Kantor, S. King, C. Bartels, M. Magimai-Doss, J. Frankel, K. Livescu, An articulatory feature-based tandem approach and factored observation modeling, in *IEEE International Conference on Acoustics, Speech and Signal Processing (ICASSP-2007)*, Honolulu, HI (2007), pp. IV-645–IV-648. https://doi.org/10.1109/ICASSP.2007.366995

30. P. Lal, S. King, Cross-lingual automatic speech recognition using Tandem features. IEEE Trans. Audio, Speech, Language Process. **21**(12), 2506–2515 (2013). https://doi.org/10.1109/TASL.2013.2277932

31. H. Hermansky, D.P. Ellis, S. Sharma, Tandem connectionist feature extraction for conventional HMM systems, in *IEEE International Conference on Acoustics, Speech and Signal Processing (ICASSP)*, vol. 3 (2000), pp. 1635–1638. https://doi.org/10.1109/ICASSP.2000.862024

32. J. Pinto, S. Garimella, M. Magimai-Doss, H. Hermansky, H. Bourlard, Analysis of MLP-based hierarchical phoneme posterior probability estimator. IEEE Trans. Audio, Speech, Language Process. **19**(2), 225–241 (2011). https://doi.org/10.1109/TASL.2010.2045943

33. J. Frankel, M. Magimai-Doss, S. King, K. Livescu, O. Cetin, Articulatory feature classifiers trained on 2000 hours of telephone speech, in *INTERSPEECH* (2007), pp. 2485–2488

34. H. Ketabdar, H. Bourlard, Hierarchical integration of phonetic and lexical knowledge in phone posterior estimation, in *IEEE International Conference on Acoustics, Speech, and Signal Processing (ICASSP)* (2008), pp. 4065–4068. https://doi.org/10.1109/ICASSP.2008.4518547

35. Sclite Tool. http://www1.icsi.berkeley.edu/Speech/docs/sctk-1.2/sclite.htm [Accessed Mar. 08, 2020]

Chapter 4
Prediction of Multilingual Articulatory Features

4.1 Introduction

In the previous chapter, the development and analysis of Multi-PRS for Indian languages is discussed. In this chapter, the prediction of multilingual AFs using four Indian languages—Kannada, Telugu, Bengali, and Odia—is described. Articulatory Feature Predictors (AF-Predictors) are developed for predicting the AFs from spectral features. The transcription derived using *IPA* chart is used for obtaining the AF labels from phone labels. MFCCs are used for representing the spectral features. DNNs are used for training the AF-predictors. AF-predictors for five AF groups—place, manner, roundness, frontness, and height—are developed. The MTL is explored to improve the prediction accuracy of AF-predictors. This chapter is organized as follows: Sect. 4.2 provides the AF specification for five AF groups used in this study. Section 4.3 describes the development and evaluation of AF-predictors. In Sect. 4.4, the use of MTL approach to improve the performance of AF-predictors is discussed. Section 4.5 summarizes the contents of this chapter.

4.2 Articulatory Features Specification

Each sound unit can be represented as a set of features based on the articulators used to produce it. These features that describe the properties of speech production of a sound unit are called AFs. The positioning and movement of various articulators involved in the production of a specific sound unit is captured using AFs. AFs vary from one sound unit to another. The co-articulation effect between the adjacent phonetic units is captured by the AFs [1–4]. The AFs act as additional clues, which aid in discriminating between various sound units. AFs provide supplementary information, which can be used along with the spectral features to improve the performance of Multi-PRSs. The AFs can be continuous or discrete [5], with the

Manjunath K. E., *Multilingual Phone Recognition in Indian Languages*,
SpringerBriefs in Speech Technology, https://doi.org/10.1007/978-3-030-80741-2_4

Table 4.1 Articulatory features specification for different AF groups of multilingual speech corpora

AF Group (Cardinality)	Feature values	Example phones	
		/p/	/i/
Place (9)	bilabial, labiodental, alveolar, retroflex, palatal, velar, glottal, vowel, silence	bilabial	vowel
Manner (6)	plosive, fricative, approximant, nasal, vowel, silence	plosive	vowel
Roundness (4)	rounded, unrounded, consonant, silence	consonant	unrounded
Frontness (5)	front, mid, back, consonant, silence	consonant	front
Height (6)	close, close-mid, open-mid, open, consonant, silence	consonant	close

Mermelstein model [6–8] being a classic example of the continuous model. Both continuous valued AFs [1, 9, 10] and discrete valued AFs [11–13] are explored in the literature to improve the performance of speech recognizers. It is found that the use of AFs has consistently improved the performance of speech recognizers.

In this work, the discrete information about the positioning and movement of articulators with respect to five AF groups is captured [14, 15]. Each AF group and their possible AF values (i.e. AF specification) for multilingual speech corpora is shown in Table 4.1. First column indicates the AF group and the cardinality. The cardinality indicates the number of feature values in an AF group. Second column lists the possible feature values for each AF group. Third and fourth columns provide the sample representation of the AF values for /p/ and /i/ phones, respectively. Similar kinds of AF specifications are reported in [11–13].

4.3 Articulatory Feature Predictors (AF-Predictors)

In this section, the development and evaluation of AF-Predictors for predicting the AFs from spectral features is described. The AF-predictors for five AF groups— place, manner, frontness, roundness, and height—are developed. The *place* and *manner* AF groups help in capturing the characteristics of consonants, while the *frontness*, *roundness*, and *height* AF groups aid in capturing the characteristics of vowels. The experimental setup such as—multilingual speech corpora, feature extraction, and procedure for training DNNs—is same as the one described in Sect. 3.2.

4.3.1 Development of Articulatory Feature Predictors

In this work, the frame-level AFs for each AF group are predicted from the spectral features using AF-predictors. Separate AF-predictors are developed for each AF group. AFs are predicted for five AF groups—place, manner, roundness, frontness, and height—using AF-predictors. The spectral features are represented using MFCCs. The AF-predictors are trained using context dependent DNNs. For training DNNs, the speech data that is transcribed using AF labels at frame-level is required. Since the transcription is available at phone-level, the frame-level AF labels for each AF group are obtained by mapping the phone labels (present in the phone-level transcription) to AF labels at frame-level. The AF label of an AF group represents a possible AF value for that specific AF group. The possible AF labels for each AF group are shown in Table 4.1. Table 4.2 shows the mapping of each phone label into a set of AF labels of various AF groups. First column in Table 4.2 lists the unique IPA symbols present in the IPA transcription of multilingual speech corpora from which the *common multilingual phone-set* is derived using the procedure described in Sect. 3.3.2. Second to sixth columns show the corresponding place, manner, roundness, frontness, and height AF values, respectively, for each phone. The mapping for each IPA symbol to various AF groups is derived using the IPA chart [16].

The AF-predictors for all AF groups are trained using the AF labelled data of four languages—KN, TE, BN, and OD. Since the AF-predictors are trained using the data of multiple languages from multilingual speech corpora, the AFs thus obtained are called *multilingual AFs*. AF-predictors are trained for classification of the features shown in Table 4.1. The posterior probabilities generated by the AF-predictors represent AFs. Figure 4.1 illustrates the prediction of manner AFs (i.e. manner AF-predictor). The predicted feature values represent the manner AFs. The block diagrams for remaining four AF-predictors are similar to Fig. 4.1. For comparison of the results, the AF-predictors are also trained using shallow neural networks (i.e. context dependent FeedForward Neural Networks (FFNNs) with 1 hidden layer) in addition to training AF-predictors using DNNs.

Figure 4.2 shows the illustration of prediction of manner AFs for ten frames using the posteriogram representation. In order to get better visualization of the posteriogram distribution across all the feature values, the posteriogram is plotted using non-consecutive frames. The labels on the X-axis of posteriogram indicate the feature values of manner AF group. The first block in the figure indicates the ground truth AF labels for each frame. MFCCs are extracted from each frame and fed to manner AF-predictor to derive the posteriogram distribution for that specific frame. The posteriogram distribution represents the manner AFs. The darker bands in the posteriogram indicate higher posterior probability, while the lighter bands indicate lower posterior probability. The sum of all the posterior probabilities obtained for a frame will be equal to 1.

Table 4.2 Mapping of phone labels in multilingual speech corpora to AF values of various AF groups

Phones	Articulatory Feature Groups				
	Place	Manner	Roundness	Frontness	Height
a æ	vowel	vowel	unrounded	front	open
ɛ	vowel	vowel	unrounded	front	open-mid
e	vowel	vowel	unrounded	front	close-mid
i ɪ	vowel	vowel	unrounded	front	close
ɑ	vowel	vowel	unrounded	back	open
u ʊ	vowel	vowel	rounded	back	close
ɔ	vowel	vowel	rounded	back	open-mid
ɐ	vowel	vowel	unrounded	mid	open
ə	vowel	vowel	unrounded	mid	open-mid
o	vowel	vowel	rounded	back	close-mid
ɘ	vowel	vowel	unrounded	mid	close-mid
ɒ	vowel	vowel	rounded	back	open
œ	vowel	vowel	rounded	front	open-mid
k kʰ g gʰ q	velar	plosive	consonant	consonant	consonant
p pʰ b bʰ	bilabial	plosive	consonant	consonant	consonant
t tʰ d dʰ	alveolar	plosive	consonant	consonant	consonant
ʧʧʰ ʤʤʰ	palatal	plosive	consonant	consonant	consonant
ʈ ʈʰ ɖ ɖʰ	reftroflex	plosive	consonant	consonant	consonant
r ɹ l	alveolar	approximant	consonant	consonant	consonant
v ʋ	labiodental	approximant	consonant	consonant	consonant
j	palatal	approximant	consonant	consonant	consonant
ɭ	reftroflex	approximant	consonant	consonant	consonant
m	bilabial	nasal	consonant	consonant	consonant
ŋ	velar	nasal	consonant	consonant	consonant
n	alveolar	nasal	consonant	consonant	consonant
ɲ	palatal	nasal	consonant	consonant	consonant
ɳ	reftroflex	nasal	consonant	consonant	consonant
s sʃʒz	alveolar	fricative	consonant	consonant	consonant
h	glottal	fricative	consonant	consonant	consonant
f	labiodental	fricative	consonant	consonant	consonant
x	velar	fricative	consonant	consonant	consonant
sil	silence	silence	silence	silence	silence

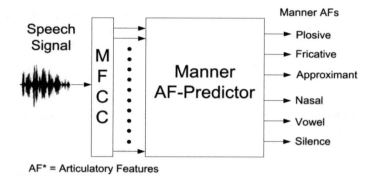

AF* = Articulatory Features

Fig. 4.1 Block diagram of Manner Articulatory Features Predictor

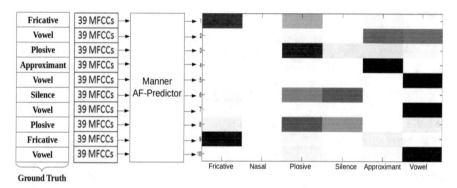

Fig. 4.2 Illustration of Prediction of Manner Articulatory Features for Ten non-consecutive frames using Posteriogram representation

Figure 4.3 illustrates the development of AF-predictors for five AF groups. The AFs for a particular AF group are predicted using the AF-predictor of that specific group.

4.3.2 Oracle Articulatory Features

In addition to the prediction of AFs through AF-predictors, the oracle AFs are also derived for better comparison and analysis. The oracle AFs for each AF group are derived from the ground truth IPA transcriptions using the following procedure. The phone labels are mapped to AF labels at frame-level using Table 4.2. The frame-level posteriogram for oracle AFs is generated by setting the posterior corresponding to the AF label to 1 and remaining posteriors to 0. The posteriogram thus generated will be used as oracle AFs. Table 4.3 provides the sample representation of oracle AFs for phones /p/ and /i/. First column shows the AF groups. Second and third

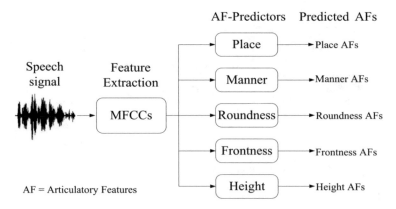

Fig. 4.3 Block diagram for the Prediction of Articulatory Features

Table 4.3 Sample representation of Oracle AFs for phones /p/ and /i/

| | Example phones | | | |
	/p/		/i/	
AF Group	AF value	Oracle AFs	AF value	Oracle AFs
Place	bilabial	(1,0,0,0,0,0,0,0,0)	vowel	(0,0,0,0,0,0,0,1,0)
Manner	plosive	(1,0,0,0,0,0)	vowel	(0,0,0,0,1,0)
Roundness	consonant	(0,0,1,0)	unrounded	(0,1,0,0)
Frontness	consonant	(0,0,0,1,0)	front	(1,0,0,0,0)
Height	consonant	(0,0,0,0,1,0)	close	(1,0,0,0,0,0)

columns indicate the AF values and oracle AFs representation for phone /p/ (consonant phone). Similarly, last two columns indicate the AF values and oracle AFs representation for phone /i/ (vowel phone).

4.3.3   Performance Evaluation of AF-Predictors

The performance of AF-predictors is evaluated using 3 measures: (i) Frame-wise accuracy, (ii) Mean Squared Error (MSE), and (iii) AF-Prediction Error Rate (AF-PER). The frame-wise accuracy is a measure of accuracy whereas the MSE and AF-PER methods measure the error rates. The label with maximum posterior value in the posteriogram of a frame will be decoded as the AF label for that frame. The frame-wise accuracy of each AF-predictor is computed by comparing the decoded AF label with the actual AF label at frame-level [12, 13]. MSE measures the average of the squares of the errors between the predicted and oracle AFs. Both the predicted and oracle AFs are represented as posterior vectors. AF-PER of AF-predictors is computed by comparing the decoded AF labels with the reference AF labels using dynamic programming. AF-PER is computed similar to the computation of PER

Table 4.4 Frame-wise accuracies, mean squared errors, and AF-prediction error rates of various AF-predictors (HL = hidden layer)

AF Group	FFNN (1 HL)	DNN (5 HL)
	Frame-wise Accuracy(%)	
Place	80.5	85.6
Manner	85.7	89.4
Roundness	87.1	90.8
Frontness	81	84.8
Height	77	80.5
	Mean Squared Error	
Place	0.033	0.025
Manner	0.038	0.028
Roundness	0.055	0.037
Frontness	0.060	0.048
Height	0.059	0.051
	AF-Prediction Error Rate(%)	
Place	26.5	21.3
Manner	20.8	17.9
Roundness	21.2	18.5
Frontness	26.0	23.3
Height	28.6	26

[17] using Eq. (3.2) as described in Sect. 3.4 except that the AF labels are used for comparison in-place of the phone labels.

Table 4.4 shows the frame-wise accuracy, MSE, and AF-PER of various AF-predictors. The results of FFNNs (1 hidden layer) and DNNs (5 hidden layers) are shown separately. It is found that the performance of DNNs is much better compared to that of FFNNs for all the AF groups. This indicates that the use of DNNs is more beneficial compared to shallow neural networks for estimating the AFs. Hence, only DNN-based AF-predictors are considered in all upcoming experiments. *Roundness* AF group shows the highest frame-wise accuracy, while the *height* AF group has shown least frame-wise accuracy. *Place* AF group shows the least MSE, while the *height* AF group has shown highest MSE. *Manner* AF group shows the least AF-PER, while the *height* AF group has shown highest AF-PER. In addition to the five AF groups considered here, the *voicing* AF group with three classes—*voiced, unvoiced, and silence*—is also tried out. But, the performance of *voicing* AF-predictor is found to be very poor. The poor performance was mainly due to the large number of misclassifications between silence and unvoiced.

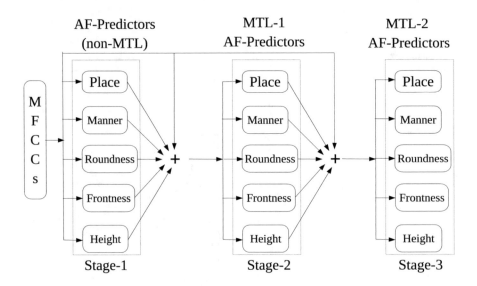

Fig. 4.4 Block Diagram of the Development of AF-predictors using Multitask Learning

4.4   Performance Improvement of AF-Predictors using Multitask Learning (MTL)

The theory of multitask learning states that by jointly learning different related tasks which share the same input and some internal representation, the performance of each task can be improved [18–20]. In this section, the use of MTL approach for joint estimation of AFs of various AF groups is investigated. The MTL approach is explored to improve the prediction accuracy of AF-predictors. The approach used for development of MTL based AF-predictors is similar to the one described in [21].

Figure 4.4 shows the block diagram of the development of AF-predictors using MTL approach. The block diagram has three stages. Stage-1 is same as the one shown in Fig. 4.3. In *stage-1*, DNNs are trained to develop AF-predictors using MFCCs as features. These AF-predictors are not based on MTL and hence, refer them as *non-MTL AF-predictors*. In *Stage-2*, MFCCs and AFs from all the five AF-predictors of *stage-1* are concatenated and used as input to train separate DNNs for each of the five AF-predictors. The AF-predictors developed in *stage-2* are called *MTL-1 AF-predictors*. Similarly, *MTL-2 AF-predictors* are developed in *stage-3* by concatenating the MFCCs and AFs from all the five AF-predictors of *stage-2* followed by training a separate DNN for each of the five AF-predictors.

Table 4.5 shows the AF-PER of various AF-predictors developed using non-MTL (stage-1) and MTL (stage-2 and stage-3) approaches. The procedure for computation of AF-PER of AF-predictors is mentioned in Sect. 4.3.3. AF-PERs of

Table 4.5 AF-prediction error rates of various AF-predictors using with and without MTL approaches

AF-predictor	AF-Prediction Error Rate(%)		
	non-MTL (Stage-1)	MTL-1 (Stage-2)	MTL-2 (Stage-3)
Place	21.3	19.6	19.5
Manner	17.9	16.7	16.5
Roundness	18.5	16.4	16.0
Frontness	23.3	20.6	20.3
Height	26	22.8	22.3

non-MTL AF-predictors are same as the ones shown in Table 4.4 and are repeated here for convenience and for better comparison with the MTL based AF-predictors. The MTL-1 and MTL-2 systems correspond to stage-2 and stage-3 blocks of Fig. 4.4, respectively. The AF-PERs of all the MTL based AF-predictors have consistently reduced compared to the non-MTL AF-predictors. All the non-MTL AF-predictors have shown the highest AF-PERs, whereas all the MTL-2 AF-predictors have shown the least AF-PERs. The AF-PERs of MTL-1 AF-predictors are in between that of non-MTL and MTL-2 AF-predictors.

4.5 Summary

AF-predictors are developed for predicting the AFs for five AF groups namely— place, manner, roundness, frontness, and height. The context dependent DNNs and shallow neural networks are explored for training AF-predictors. The performance of AF-predictors is evaluated using 3 measures: (i) frame-wise accuracy, (ii) mean squared error, and (iii) AF-prediction error rate. The DNN-based AF-predictors have shown better performance compared to AF-predictors using shallow neural networks. It is observed that the MTL based AF-predictors have better performance compared to non-MTL AF-predictors. The MTL-2 AF-predictors have shown least AF-PERs for all the AF groups. The use of MTL has improved the prediction accuracy of AF-predictors (i.e. reduced AF-PER of AF-predictors).

References

1. V. Mitra, W. Wang, A. Stolcke, H. Nam, C. Richey, J. Yuan, M. Liberman, Articulatory trajectories for large-vocabulary speech recognition, in *IEEE International Conference on Acoustics, Speech and Signal Processing (ICASSP)* (2013), pp. 7145–7149. https://doi.org/10.1109/ICASSP.2013.6639049
2. K. Erler G.H. Freeman, An HMM-based speech recognizer using overlapping articulatory features. J. Acoustic Soc. Am. **100**(4), 2500–2513 (1996). https://doi.org/10.1121/1.417358

3. S.E.G. Ohman, Coarticulation in VCV utterances: spectrographic measurements. J. Acoustic Soc. Am. **39**(1), 151–68 (1965). https://doi.org/10.1121/1.1909864
4. V.R. Ramachandran, Coarticulation knowledge for a Text-to-speech system for an Indian language, MS Thesis (Indian Institute of Technology Madras, Chennai, 1993)
5. S. King, J. Frankel, K. Livescu, E. McDermott, K. Richmond, M. Wester, Speech production knowledge in automatic speech recognition. J. Acoustic Soc. Am. **121**(2), 723–724 (2007). https://doi.org/10.1121/1.2404622
6. P. Mermelstein, Computer simulation of articulatory activity in speech production, in *Proceedings of International Joint Conference on Artificial Intelligence (IJCAI)* (1969), pp. 447–454
7. P. Mermelstein, Articulatory model for the study of speech production. J. Acoust. Soc. Am. **53**(4), 1070–1082 (1973). https://doi.org/10.1121/1.2404622
8. P. Rubin, T. Baer, P. Mermelstein, An articulatory synthesizer for perceptual research. J. Acoust. Soc. Am. **70**(2), 321–328 (1981). https://doi.org/10.1121/1.386780
9. J. Frankel, S. King, Speech recognition using linear dynamic models. IEEE TASLP **15**(1), 246–256 (2007). https://doi.org/10.1109/TASL.2006.876766
10. I. Zlokarnik, Adding articulatory features to acoustic features for automatic speech recognition. J. Acoust. Soc. Am. **97**(5), 3246–3246 (1995). https://doi.org/10.1121/1.411699
11. K. Kirchhoff, G.A. Fink, G. Sagerer, Combining acoustic and articulatory feature information for robust speech recognition. Speech Commun. **37**, 303–319 (2002). https://doi.org/10.1016/S0167--6393(01)
12. J. Frankel, M. Magimai-Doss, S. King, K. Livescu, O. Cetin, Articulatory feature classifiers trained on 2000 hours of telephone speech, in *INTERSPEECH* (2007), pp. 2485–2488
13. O. Cetin, A. Kantor, S. King, C. Bartels, Magimai-Doss, J. Frankel, K. Livescu, An articulatory feature-based tandem approach and factored observation modeling, in *IEEE International Conference on Acoustics, Speech and Signal Processing (ICASSP-2007)* , Honolulu, HI (2007), pp. IV-645–IV-648. https://doi.org/10.1109/ICASSP.2007.366995
14. K.E. Manjunath, K.S. Rao, D.B. Jayagopi, V. Ramasubramanian, Indian languages ASR: A multilingual phone recognition framework with IPA based common phone-set, predicted articulatory features and feature fusion, in *INTERSPEECH*, Hyderabad (2018), pp. 1016–1020. https://doi.org/10.21437/Interspeech.2018--2529
15. K.E. Manjunath, D.B. Jayagopi, K.S. Rao, V. Ramasubramanian, *Articulatory Feature based Methods for Performance Improvement of Multilingual Phone Recognition Systems using Indian Languages.* Sadhana (Springer) (2020)
16. The International Phonetic Association, *Handbook of the International Phonetic Association* (Cambridge University, Cambridge, 2007). https://www.internationalphoneticassociation.org/ [Accessed Mar. 08, 2020]
17. Sclite Tool. http://www1.icsi.berkeley.edu/Speech/docs/sctk-1.2/sclite.htm [Accessed Mar. 08, 2020]
18. R. Caruana, Multitask learning, in *Learning to Learn*, ed. by S. Thrun, L. Pratt (Springer, Berlin, 1998), pp. 95–133. https://doi.org/10.1023/A:1007379606734
19. R. Sahraeian, Acoustic Modeling of Under-resourced Languages, Ph.D. Thesis (Katholieke Universiteit Leuven (KU Leuven), Leuven, 2017)
20. M. Muller, S. Stuker, A. Waibel, Towards improving low-resource speech recognition using articulatory and language features, in *International Workshop on Spoken Language Translation (IWSLT)* (2016), pp. 1–7
21. R. Rasipuram, M. Magimai-Doss, Improving articulatory feature and phoneme recognition using multitask learning, in *Artificial Neural Networks and Machine Learning (ICANN)*, vol. 6791 (2011), pp. 299–306. https://doi.org/10.1007/978-3-642-21735-7_37

Chapter 5
Articulatory Features for Multilingual Phone Recognition

5.1 Introduction

In the previous chapter, the prediction of multilingual AFs using Articulatory Feature Predictors (AF-Predictors) is discussed. In this chapter, the use of predicted AFs to improve the performance of Multi-PRS is described. Both HMMs and DNNs are explored for training Multi-PRS. The oracle AFs, which are derived from the ground truth IPA transcriptions, are used to set the best performance realizable by the predicted AFs. The performance of Multi-PRSs using the predicted and oracle AFs is compared. In addition to the AFs, the phone posteriors are explored to further boost the performance of Multi-PRS. Fusion of AFs from different AF groups is done using two approaches: (i) lattice rescoring approach, (ii) AFs as tandem features. The AFs obtained from the MTL approach are examined to further improve the performance of Multi-PRS. This chapter is organized as follows: Sect. 5.2 describes the proposed approaches for multilingual phone recognition using AFs. Section 5.3 discusses the development of Multi-PRSs using MTL based AFs. Section 5.4 summarizes the contents of this chapter.

5.2 Proposed Approaches for Multilingual Phone Recognition using Articulatory Features

In this section, two approaches for multilingual phone recognition using AFs are briefly outlined. Based on the way in which the predicted AFs from different AF groups are combined, two approaches—namely (i) Lattice Rescoring Approach (LRA), (ii) Combining AFs as Tandem features (AF-Tandem)—are proposed. AF-predictors are trained using DNNs to predict the AFs from MFCCs as described in Chap. 4. The AFs predicted from the AF-predictors are called predicted AFs. The predicted AFs and MFCCs are combined to improve the performance of Multi-

© The Author(s), under exclusive license to Springer Nature Switzerland AG 2022   57
Manjunath K. E., *Multilingual Phone Recognition in Indian Languages*,
SpringerBriefs in Speech Technology, https://doi.org/10.1007/978-3-030-80741-2_5

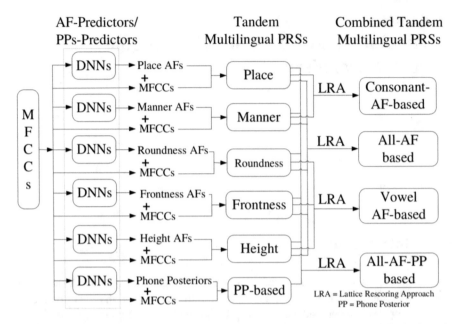

Fig. 5.1 Block diagram of Proposed Multi-PRS using Lattice Rescoring Approach for Fusion of Articulatory Features

PRSs. In addition to AFs, the combination of PPs is also explored to further boost the performance of Multi-PRS.

Figure 5.1 shows the block diagram of combination of AFs using LRA.

There are 3 stages in Fig. 5.1. In the first stage, the AF-predictors are developed to predict the AFs for five AF groups from MFCCs. DNNs are used to develop AF-predictors. In the second stage, the predicted AFs (output of first stage) are combined with the MFCCs to develop Multi-PRSs. Since these Multi-PRSs are developed using AFs and are arranged in tandem, let us refer them AF-based tandem Multi-PRSs. The third stage is developed to combine the AFs from multiple AF groups. In the third stage, LRA is used for combining the AF-based tandem Multi-PRSs developed in the second stage [1, 2].

In AF-tandem approach for fusion AFs, the predicted AFs from different AF-predictors are augmented with MFCCs and used as tandem features to develop Multi-PRSs. Figure 5.2 shows the block diagram of proposed Multi-PRS that uses AF-Tandem approach for combining the AFs of different AF groups.

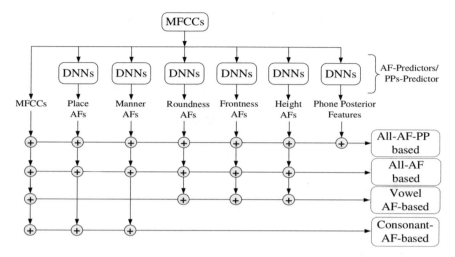

Fig. 5.2 Block diagram of Proposed Multi-PRS using AF-Tandem Approach for Fusion of Articulatory Features

Table 5.1 PERs of baseline multilingual phone recognition systems developed using MFCCs

	Context independent		Context dependent	
Baseline	HMM	DNN	HMM	DNN
Multi-PRS	49.4	39.8	39.0	35.1

5.2.1 Development of AF-Based Tandem Multilingual Phone Recognition Systems

Multi-PRS is developed using the data of four Indian languages—Kannada, Telugu, Bengali, and Odia—as described in Sect. 3.3.2. The baseline Multi-PRS is developed using MFCC features. Table 5.1 shows the PERs of baseline Multi-PRS. The results shown in Table 5.1 are same as the results shown for the *KN-TE-BN-OD* Multi-PRS (i.e. Quadri-PRS) in Table 3.5. They are repeated here for convenience and better comparison. It is observed that the baseline Multi-PRS using context dependent DNNs has a best PER of 35.1%.

AF-based tandem Multi-PRSs are developed using the combination of MFCCs and the predicted AFs. The AFs for each AF group are predicted from the spectral features using the AF-predictors as described in Chap. 4. In tandem approach, the DNNs are first trained using MFCCs to perform the classification at frame-level, and then the frame-level posterior probability estimates of the DNNs are used as features for developing Multi-PRSs. The predicted AFs of a particular AF group are augmented with the MFCCs to develop AF-based tandem Multi-PRS for that AF group [3–5]. Separate tandem Multi-PRSs are developed using the AFs predicted from each AF group. Five AF-based tandem Multi-PRSs are developed by using the combination of MFCCs and the AFs predicted by the corresponding AF-predictor.

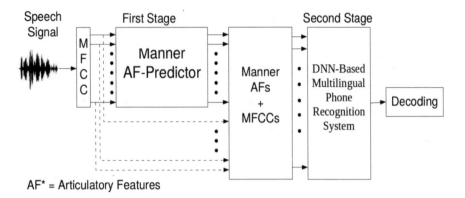

Fig. 5.3 Block diagram of the Manner AF-based Tandem Multilingual Phone Recognition System

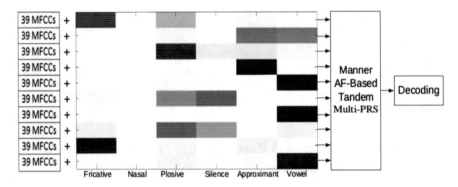

Fig. 5.4 Illustration of Manner AF-based Tandem Multi-PRS for Ten frames using posteriogram representation

Figure 5.3 shows the block diagram of manner AF-based tandem Multi-PRS. The manner AF-predictor is used for predicting the manner AFs as shown in Fig. 4.1. The combination of predicted manner AFs and MFCCs are used as features for training DNNs to develop a manner AF-based tandem Multi-PRS in the second stage. Similarly, five different AF-based tandem Multi-PRSs are developed using the predicted AFs from each AF group.

Figure 5.4 illustrates the Manner AF-based tandem PRS for ten frames using posteriogram representation. The MFCCs are augmented with the posteriogram distribution of the manner AFs obtained in first stage (shown in Fig. 4.2). The combination of the MFCCs and the manner AFs is then fed to the manner AF-based tandem Multi-PRS for decoding the phones in the input speech utterance.

The oracle AFs are used to establish the best target performance achievable by the predicted AFs of each AF group. The oracle AFs are obtained by using the procedure mentioned in Sect. 4.3.2. Table 5.2 shows the PERs of AF-based tandem

Table 5.2 Phone error rates of AF-based tandem multilingual phone recognition systems

Type of AFs	Features	PER (%) of CD DNNs	
		Predicted AFs	Oracle AFs
Consonant AFs	MFCCs + Place	33.5	21.1
	MFCCs + Manner	34.1	24.0
Vowel AFs	MFCCs + Roundness	34.9	26.8
	MFCCs + Front	34.1	26.9
	MFCCs + Height	34.3	23.1

Multi-PRSs. First column shows the type of AFs (i.e. consonant or vowel AFs). Second column indicates the features used for building various AF-based tandem Multi-PRSs. Third and fourth columns show the PERs obtained using predicted and oracle AFs, respectively. The results are shown separately for predicted and oracle AFs.

It is observed that the PERs of all the AF-based tandem Multi-PRSs are superior compared to the PER of baseline Multi-PRS (35.1%). This clearly indicates that the use of AFs has reduced the PERs. The *place* AF-based tandem Multi-PRS has shown the highest reduction in PER, and *roundness* AF-based tandem system has shown least reduction using predicted AFs. This is because, *place* AF group is a very important distinctive feature [6, 7]. Hence, it gives the *class* information more readily than other AF groups, with manner AF group being the next best. It can also be noted that the *place* AF group has highest cardinality (i.e. 9), while the *roundness* has least cardinality (i.e. 4) as shown in Table 4.1. The cardinality indicates number of feature classes (i.e. feature dimension). Higher cardinality (higher feature dimension) provides more discriminative information to classify among various phonetic units. This results in improved phone recognition accuracy and reduces the PER. Similarly, lower cardinality would lead to higher PER.

The consonant-AF-based systems (see row 2,3 of Table 5.2) have lower PERs compared to vowel-AF-based systems (see row 4,5,6 of Table 5.2). It is found that misclassifications among the consonants are reduced in consonant-AF-based systems, and the misclassifications among the vowels are reduced in vowel-AF-based systems. The average PER of oracle AFs is 14.5% lower than that of predicted AFs. This indicates that there is large scope to reduce the PERs of predicted AFs (up to 14.5% on an average). Hence, the simultaneous combination of AFs from various AF groups is further explored.

5.2.2   Fusion of AFs from Multiple AF Groups

The AFs from different AF groups are combined together to take the mutual advantage of all the AFs at the same time. Two approaches, namely (i) LRA approach, (ii) AF-Tandem approach, are explored for fusion of AFs from different

Table 5.3 Phone error rates of combined tandem multilingual phone recognition systems

Combined Multi-PRSs	Predicted AFs		Oracle AFs	
	LRA	AF-Tandem	LRA	AF-Tandem
Vowel-AF-based	33.4	34.8	22.1	21.8
Consonant-AF-based	33.0	33.7	19.6	17.8
All-AF-based	32.7	33.5	12.9	10.4
All-AF-PP-based	32.6	32.3	-	-

AF groups. In LRA approach, the lattices generated by the AF-based tandem systems are combined using the lattice rescoring method [8]. The weighting factors required for LRA are tuned using development set. In AF-Tandem method of combination, the AFs are augmented as tandem features along with MFCCs to develop Multi-PRSs [3, 9]. As shown in Figs. 5.1 and 5.2, the AFs derived from the consonant AF groups are combined to develop consonant-AF-based Multi-PRS, while the vowel-AF-based Multi-PRS is developed by combining the AFs from vowel AF groups. All-AF-based Multi-PRS (see Figs. 5.1 and 5.2) is developed by combining all the five AF-based tandem systems. Further, the combining of PPs along with all the predicted AFs to develop All-AF-PP-based Multi-PRS is also explored (see Figs. 5.1 and 5.2) [10]. Similar to AFs, the PPs are predicted from the MFCCs by training a DNN [11, 12]. The PPs are predicted using the procedure described in Sect. 3.3.3.

Table 5.3 shows the PERs of different AF-based Multi-PRSs combined using LRA and AF-Tandem approaches. The results are shown separately for predicted and oracle AFs. The improvements in the performance are consistent across both predicted and oracle AFs. The Consonant-AF-based has higher PER reduction compared to Vowel-AF-based, while the All-AF-based has higher PER reduction compared to Consonant-AF-based system. The PER of All-AF-based Multi-PRS using oracle AFs is 22.3% lower than that of predicted AFs. Given the remarkably low PER of 10.4% for oracle based Multi-PRS, there is much scope for enhanced prediction of AFs so as to improve the performance of Multi-PRS to reach that of oracle AFs. Hence, there is a need to explore upon other alternative methods (including the continuous valued AFs) for more accurate prediction of AFs.

In case of predicted AFs, it is observed that the LRA method of combination has shown least PERs for consonant-AF-based, vowel-AF-based, and All-AF-based Multi-PRSs, while the AF-Tandem method of combination has shown least PER for All-AF-PP-based Multi-PRS. However, in case of Multi-PRSs using oracle AFs, the AF-Tandem method of combination has resulted in least PERs for all combinations. Since the oracle PPs are same as the ground truth reference labels, it does not make any sense to use oracle PPs as the features. Hence, none of the experiments related to All-AF-PP-based Multi-PRS is conducted using oracle PPs and the corresponding values in Table 5.3 are represented as '-' (hyphen) indicating *not applicable*. In case of oracle AFs, a drastic reduction in the PER of All-AF-based Multi-PRS is observed compared to the PERs of consonant-AF-based and vowel-AF-based Multi-PRSs.

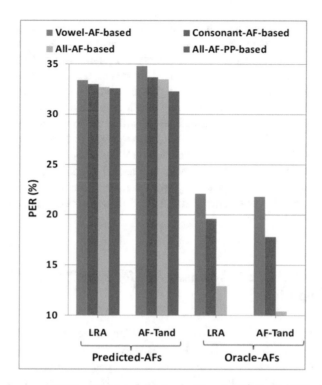

Fig. 5.5 Comparison of PERs of Predicted and Oracle AFs using LRA and AF-Tandem Methods of Combination

The All-AF-PP-based Multi-PRS using AF-Tandem method has shown the least PER of 32.3% with an absolute PER reduction of 2.8% (i.e. 8% relative PER reduction) compared to the baseline Multi-PRS. The PERs of All-AF-PP-based Multi-PRSs are superior compared to the average of the PERs of all monolingual PRSs (33.0%). The AF-Tandem method not only performs better than LRA but also has less complex structure than LRA. The time complexity of LRA is almost 5× higher than AF-Tandem in terms of both training and decoding.

Figure 5.5 shows the comparison of PERs obtained from predicted and oracle AFs using LRA and AF-Tandem approaches for combination of AFs from multiple AF groups. There are no bars corresponding the All-AF-PP-based systems using oracle AFs, since such systems do not make any sense. A bar with least height can be seen for oracle AFs using AF-Tandem approach which represents the best realizable performance of oracle AFs. Highest peak can be seen by predicted AFs with AF-tandem method for vowel-AF-based system. Consistency in PER reduction of *Vowel-AF-based, Consonant-AF-based,* and *All-AF-based* systems can be clearly observed across all the systems of predicted and oracle AFs using AF-Tandem and LRA methods.

There are 33 consonants and 11 vowels in the phone-set considered. Around 55% of the test data is made of consonants, whereas only 45% constitutes vowels. Out of 45% of vowel data 15% is wrongly classified, while 26% out of 55% of consonant data is wrongly classified. This means that there is a larger scope to reduce the misclassifications within the consonants than vowels. Since the consonant AFs mainly reduce the misclassifications within the consonants and there is larger scope to reduce the misclassifications within the consonants, the consonant-AF-based Multi-PRS has shown higher improvement in PERs compared to the vowel-AF-based Multi-PRS. Since there are only few vowel classes and the vowels classification using MFCCs itself provides a reasonably good recognition accuracy, there is not much scope to further improve the recognition accuracies using vowel AFs. Also, the number of discriminative feature classes in consonants AFs is higher than that of vowel AFs.

5.3 Multitask Learning Based AFs for Multilingual Phone Recognition

The AFs predicted from the MTL based AF-predictors are combined to improve the performance of Multi-PRSs. The MTL based AF-predictors are developed using the procedure described in Sect. 4.4. Since the AF-predictors based on MTL-2 have least AF-PERs (see Table 4.5), only the AFs derived from MTL-2 based AF-predictors are considered for fusion of AFs. From Table 5.3, it is observed that the All-AF-PP-based Multi-PRS using AF-Tandem method has shown least PER (i.e. 32.3%) compared to all other Multi-PRSs. Hence, the MTL-2 based AFs and PPs are combined using AF-tandem method to develop All-AF-PP-based Multi-PRS.

Table 5.4 shows the PERs of combined Multi-PRSs using MTL and non-MTL based AFs. The results shown in second column correspond to non-MTL based Multi-PRSs that are described in Sect. 5.2.2. The values in the second column are taken from Table 5.3 (see AF-Tandem case shown in third column) and are repeated here for convenience and better comparison. The MTL-2 based *All-AF-PP-based* Multi-PRS has shown an improvement of 0.4% PER compared to non-MTL systems. The PER of 31.9% shown by MTL-2 based *All-AF-PP-based* Multi-PRS results in an absolute PER reduction of 3.2% (i.e. 9.1% relative PER reduction) compared to the baseline Multi-PRS.

Table 5.4 Phone error rates of combined multi-PRSs using both MTL and non-MTL based AF-predictors

Combined Multi-PRS	Phone error rate(%)	
	non-MTL	MTL-2
All-AF-PP-based	32.3	31.9

5.4 Summary

The use of AFs has improved the performance of Multi-PRSs. The combination of AFs using AF-Tandem method performs better than that of LRA method. The MTL based AFs have better performance compared to non-MTL based AFs. The fusion of *MTL-2* based AFs and phone posteriors using tandem approach has resulted in least PER. It is found that oracle AFs by feature fusion with MFCCs offer a remarkably low target PER of 10.4%, which is 24.7% absolute reduction compared to baseline Multi-PRS with MFCCs alone. The best performing system using predicted AFs has shown 3.2% reduction in absolute PER (9.1% reduction in relative PER) compared to baseline Multi-PRS. Given the remarkably low PER of 10.4% for oracle based Multi-PRS, it is concluded that there is much scope for enhanced prediction of AFs to improve the Multi-PRS to reach the performance of oracle AFs.

References

1. K.E. Manjunath, K.S. Rao, D.B. Jayagopi, V. Ramasubramanian, Indian languages ASR: a multilingual phone recognition framework with IPA based common phone-set, predicted articulatory features and feature fusion in *INTERSPEECH*, Hyderabad (2018), pp. 1016–1020. https://doi.org/10.21437/Interspeech.2018-2529
2. K.E. Manjunath, D.B. Jayagopi, K.S. Rao, V. Ramasubramanian, *Articulatory Feature based Methods for Performance Improvement of Multilingual Phone Recognition Systems using Indian Languages*. Sadhana (Springer) (2020)
3. O. Cetin, A. Kantor, S. King, C. Bartels, Magimai-Doss, J. Frankel, K. Livescu, An articulatory feature-based tandem approach and factored observation modeling, in *IEEE International Conference on Acoustics, Speech and Signal Processing (ICASSP-2007)*, Honolulu, HI (2007), pp. IV-645–IV-648. https://doi.org/10.1109/ICASSP.2007.366995
4. H. Hermansky, D.P. Ellis, S. Sharma, Tandem Connectionist Feature Extraction for Conventional HMM Systems, in *IEEE International Conference on Acoustics, Speech and Signal Processing (ICASSP)*, vol. 3 (2000), pp. 1635–1638. https://doi.org/10.1109/ICASSP.2000.862024
5. P. Lal, S. King, Cross-lingual automatic speech recognition using tandem features. IEEE Trans. Audio Speech Lang. Process. **21**(12), 2506–2515 (2013). https://doi.org/10.1109/TASL.2013.2277932
6. K.N. Stevens, Toward a model for lexical access based on acoustic landmarks and distinctive features. J. Acoust. Soc. Am. **111**(4), 1872–1891 (2002). https://doi.org/10.1121/1.1458026
7. K.N. Stevens, H.M. Hanson, Articulatory-acoustic relations as the basis of distinctive contrasts, in *The Handbook of Phonetic Sciences*, 2 edn., ed. by W.J. Hardcastle, J. Laver, F.E. Gibbon (Blackwell Publishing, New York, 2010), pp. 424–453. https://doi.org/10.1002/9781444317251.ch12
8. S.M. Siniscalchi, J. Li, C. Lee, A study on lattice rescoring with knowledge scores for automatic speech recognition, in *INTERSPEECH* (2006), pp. 517–520
9. J. Frankel, M. Magimai-Doss, S. King, K. Livescu, O. Cetin, Articulatory feature classifiers trained on 2000 hours of telephone speech, in *INTERSPEECH* (2007), pp. 2485–2488
10. R. Rasipuram, M. Magimai-Doss, Integrating articulatory features using Kullback-Leibler divergence based acoustic model for phoneme recognition, in *IEEE International Conference on Acoustics, Speech, and Signal Processing (ICASSP)* (2011), pp. 5192–5195. https://doi.org/10.1109/ICASSP.2011.5947527

11. H. Ketabdar, H. Bourlard, Hierarchical Integration of Phonetic and Lexical Knowledge in Phone Posterior Estimation, in *IEEE International Conference on Acoustics, Speech, and Signal Processing (ICASSP)* (2008), pp. 4065–4068
12. H. Ketabdar, H. Bourlard, Enhanced phone posteriors for improving speech recognition systems. IEEE Trans. Audio, Speech, Language Process. **18**(6), 1094–1106 (2010). https://doi.org/10.1109/TASL.2009.2023162

Chapter 6
Applications of Multilingual Phone Recognition in Code-Switched and Non-code-Switched Scenarios

6.1 Introduction

In the previous chapter, use of multilingual AFs to improve the performance of multilingual phone recognizers is discussed. In this chapter, two different approaches for multilingual phone recognition using non-code-switched and code-switched scenarios are evaluated and compared. The first approach is LID-Mono approach where a front-end LID-switched to a monolingual phone recognizer trained individually on each of the languages present in multilingual dataset. In the second approach, the *common multilingual phone-set* derived from the IPA transcription of the multilingual dataset is used to develop a Multi-PRS. *Throughout this chapter the LID-switched monolingual approach (i.e. first approach) is denoted as LID-Mono, and the common multilingual phone-set approach (i.e. second approach) is denoted as Multi-PRS.* The bilingual code-switching experiments are conducted using Kannada and Urdu languages. In the first approach, LID is performed using the state-of-the-art i-vectors. Both monolingual and multilingual phone recognition systems are trained using DNNs. The performance of LID-Mono and Multi-PRS approaches are compared and analysed in detail. This chapter is organized as follows: Sect. 6.2 describes the experimental setup used in this study. In Sect. 6.3, two approaches for multilingual phone recognition are discussed. Section 6.4 describes the performance evaluation and comparison of the two multilingual phone recognition approaches. Section 6.5 summarizes the contents of this chapter.

6.2 Experimental Setup

In this chapter, DNNs and HMMs are explored for training the PRS. The procedure used for training the GMM-HMMs and DNN-HMMs is same as the one described

Manjunath K. E., *Multilingual Phone Recognition in Indian Languages*, SpringerBriefs in Speech Technology, https://doi.org/10.1007/978-3-030-80741-2_6

Table 6.1 Statistics of Urdu and Assamese speech corpora

Language	# Speakers		Duration (in hours)			
	Male	Female	Train	Dev	Test	Total
Urdu	53	6	4.12	0.46	1.04	5.64
Assamese	8	8	2.39	0.23	0.53	2.39

Sect. 3.2.3. MFCCs are used as features for building LID systems and PRSs. The MFCCs are extracted as per the procedure mentioned in Sect. 3.2.2. The experimental setup, such as multilingual speech corpora, code-switched test set, training of Support Vector Machines (SVMs), and extraction of i-vectors, that are specific to this chapter are described in the following subsections.

6.2.1 Multilingual Speech Corpora

In addition to the *four* Indian languages (i.e. Kannada, Telugu, Bengali, and Odia) described in Sect. 3.2.1 of Chap. 3, *two* more Indian languages—namely Urdu and Assamese—are also considered in this chapter. The details of *four* Indian languages—Kannada, Telugu, Bengali, and Odia—are same as described in Sect. 3.2.1. The specific details of Urdu and Assamese speech corpora are given in Table 6.1. First column lists the names of the languages. Next two columns provide the count of number of male and female speakers, while the fourth to seventh columns tabulate the duration of different datasets in terms of number of hours.

6.2.2 Code-Switched Test Set

Mixing of *Hindi or Indian English or Urdu* with some other Indian language is the most common case of bilingual code-switching in the context of Indian languages. Among Hindi, Indian English, and Urdu languages, the IPA transcribed speech data was available only for Urdu and not for Hindi or Indian English. Hence, in this study, the intra-sentential code-switching between Kannada and Urdu languages, with Kannada sentence being the primary language within which switching occurs to Urdu words and phrases, is considered. Bilingual KN/UR code-switching is mostly found in the Indian state of Karnataka, as the bilingual KN/UR speakers mostly reside in this Indian state of Karnataka.

The details of the test set used for testing the performance in code-switched scenario are as follows. 320 code-switched sentences having Kannada as the primary language that are code-switched to Urdu words and phrases are selected. The sentences are carefully chosen to cover all the phonetic units of Kannada and Urdu languages. *Four* male and *four* female speakers, who are bilinguals of Kannada and Urdu, are made to read 40 sentences each. The speakers are proficient in both spoken Kannada and Urdu and tend to produce KN-UR code-switched

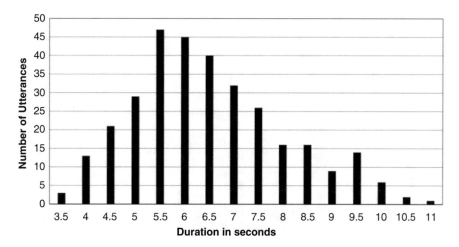

Fig. 6.1 Distribution of durations of utterances in code-switched test set

utterances. These sentences were transcribed using IPA symbols and then mapped
to the common multilingual phone-set to generate the ground-truth transcription for
calculation of PER after the decoding.

Figure 6.1 shows the histogram of distribution of durations of utterances present
in code-switched test set. The duration of each utterance ranges from 3.5 s to 11 s.
The large part of test set speech corpora lies in the range of 5.5–7.0 s. There are very
less number of sentences having their duration either greater than 10.0 s or lesser
than 4.0 s.

Figure 6.2 shows the distribution of durations of KN and UR languages in code-
switched test sets. The duration of each utterance ranges from 3.5 s to 11.0 s in
the dataset, within which the Urdu words and phrases occur at durations, from
which it can be noted that *Kannada* segments in an utterance are the longer
ones, interspersed with *Urdu* segments of relatively shorter durations, importantly
ranging from $0 - 0.5$ s to $1 - 3.5$ s which typically correspond to short words
(<500 ms) and multi-word phrases (of the order of 1–3.5 s). There are no *Kannada*
phrases in the range of 0-0.5 s indicating that the duration of all *Kannada* phrases
is >500 ms. Similarly, there are no *Urdu* phrases having a duration greater than 3.5
s. This kind of *Urdu* segments in the code-switched data plays an important factor
in determining how the LID-Mono works, particularly in the choice of the length of
speech segment on which the front-end LID has to operate.

6.2.3   Training Support Vector Machines (SVMs)

SVMs are supervised learning models associated with learning algorithms that
analyse data and recognize patterns. SVMs are used for classification and regression

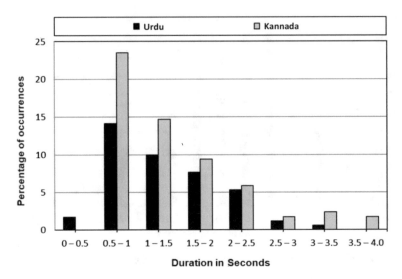

Fig. 6.2 Distribution of durations of Kannada and Urdu languages in code-switched test sets

analysis. SVMs can be used as binary linear classifiers or Multi-class non-linear classifiers. Multi-class SVMs can be constructed in several approaches including one-against-one, one-against-all. One-against-one (also called Max-win voting) approach is employed in this study. This strategy builds one SVM for each pair of classes. In this approach, if N is the number of classes, then $N(N-1)/2$ classifiers are constructed and each one trains data from two classes. The test data is applied to all machines, each classification gives one vote to the winning class, and the data is labeled with the class having most votes. If there are two identical votes, Max-win selects the class with the smallest index. The radial basis function is used as a kernel [1, 2]. SVM is implemented using *scikit-learn* machine learning tool [3].

6.2.4 Extraction of i-vectors

The i-vectors are fixed dimension features derived from variable length sequences of front-end features using a background model. The i-vectors are extensively used features for LID task and were first explored in [4]. The long term information related to language is well captured by i-vectors. GMM based Universal Background Model (UBM) is considered for i-vector modelling using Bottleneck Features (BNF) as front-end features [5]. The BNF features are extracted using the following procedure. A DNN is trained for ASR using the labelled speech data from Switchboard (i.e. SWB1 corpus) and Fisher corpora (about 2000 h). Training uses hidden layers with Rectified Linear Unit (ReLU) activation with layer-wise batch normalization. Mel-frequency Cepstral Coefficients (MFCC) are extracted from each input utterance and fed to DNNs. The BNF features (80 dimension) are extracted from the bottleneck layer of the trained DNN [6, 7]. These extracted BNF features are used as front-end features for extraction of i-vectors.

The front-end features of all the utterances of the LID training dataset are pooled together to get Gaussian Mixture Universal Background Model (GMM-UBM). The Baum-Welch (BW) statistics of the front-end features is used to adapt the means of the GMM to each utterance. The zeroth and centred first order BW statistics of a recording s and UBM mixture component c are given by the following equations.

$$N_c(s) = \sum_{i=1}^{H(s)} P(c|x_i), \qquad F_{X,c}(s) = \sum_{i=1}^{H(s)} P(c|x_i)(x_i - \mu_c)$$

where x_i is the front-end feature at time index i, F indicates the dimension of the front-end features, C is the number of mixture components of GMM-UBM, $H(s)$ indicates the number of frames in recording s, and μ_c is the mean of the UBM mixture component c.

A Total Variability Model (TVM) is assumed as a generative model for the adapted GMM mean supervector and it is represented as,

$$M(s) = M_0 + Ty(s),$$

where $M_0 = (\mu_1, \ldots \ldots \mu_C)^T$ is the UBM mean supervector, and $M(s) = (\mu_1(s), \ldots \ldots \mu_C(s))^T$ is the adapted mean supervector of a recording s. T is a matrix of dimension $CF \times R$, and $y(s) \sim N(0, I)$ is a latent variable and is a vector of dimension R. The subspace spanned by $y(s)$ is called the *total variability space*. The *maximum aposteriori* (MAP) estimate of $y(s)$ given the front-end features of the recording s is called the i-vector $y^*(s)$ of the recording s. The i-vectors thus extracted are subjected to various normalization and dimensionality reduction techniques before using them as features for the LID task. This procedure for extraction of i-vectors used in this study is similar to the one described in [8].

6.3 Approaches for Multilingual Phone Recognition

This section briefly describes two important approaches for multilingual phone recognition: (i) LID-switched monolingual phone recognition, and (ii) Multilingual phone recognition using common multilingual phone-set [9, 10].

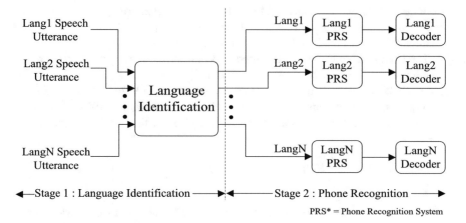

Fig. 6.3 Multilingual Phone Recognition using LID-Mono Approach

6.3.1   LID-switched Monolingual Phone Recognition (LID-Mono) Approach

LID-Mono is a traditional approach for multilingual phone recognition and is shown in Fig. 6.3. Figure 6.3 consists of two stages. In the *first stage*, the language of the input speech is determined using a language identification block. In the *second stage*, the input speech utterance is routed to the monolingual PRS of the language identified in stage-1 and the phones present in the input speech are determined. Monolingual phone recognizer is a conventional PRS developed using the data of single language. The following subsections provide the detailed description of LID block (i.e. first stage) and Mono-PRS block (i.e. second stage). Few notable prior works based on LID-Mono approach are reported in [11–13] and are briefly described in Sect. 2.2.

6.3.1.1   Development of Language Identification (LID) System

The LID system is briefly outlined here. There are two approaches for LID, namely implicit LID and explicit LID. The explicit LID requires phonetic transcription and language models for each language [14–16], whereas the implicit LID does not need either phonetic transcription or language models [17, 18]. Since the language models for the languages considered in this study were not available to us, the implicit LID is used to perform LID in this study. Both MFCCs and i-vectors are explored as features for building LID systems using SVMs. MFCCs and i-vectors are extracted as per the procedure described in Sects. 3.2.2 and 6.2.4. SVMs are used for training the LID classifiers [19–21] using the procedure described in Sect. 6.2.3. The test sets of the multilingual speech corpora (described in Tables 3.1 and 6.1) are used

Table 6.2 LID accuracy for various language sets using test sets of multilingual speech corpora

Languages	LID Accuracy (%)	
	MFCCs	i-vectors
KN-BN-OD-UR	91.16	97.98
KN-TE-BN-OD-UR	74.76	96.22
KN-TE-BN-OD-UR-AS	71.19	96.00

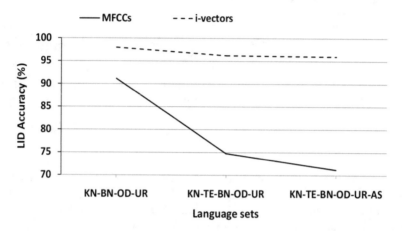

Fig. 6.4 Comparison of LID accuracies obtained from MFCCs and i-vectors for Various Language Sets

for evaluating the LID systems. The test sets of all the languages involved in building a LID system will be used for evaluating that specific LID system. The duration of each utterance in the test set varies from 3 s to 11 s.

Table 6.2 shows the LID accuracy for various language sets. First column indicates the language sets. Second and third columns show the LID accuracy using MFCCs and i-vectors, respectively.

Figure 6.4 shows the comparison of LID accuracies obtained by using MFCCs and i-vectors for various language sets. X-axis denotes the language sets, and Y-axis denotes the LID accuracy.

In all the cases, the LID accuracy of i-vectors has outperformed MFCCs. It is found that the LID accuracy decreases as the number of languages increases. Since the performance of LID using i-vectors outperforms MFCCs, only i-vector based LID systems are considered in all the upcoming experiments.

6.3.1.2   Development of Monolingual Phone Recognition Systems (Mono-PRS)

The experimental setup and the procedure used for developing the Mono-PRSs are same as described in Sect. 3.3.1. Mono-PRSs are developed for Kannada, Telugu, Bengali, Odia, Urdu, and Assamese languages using 36, 35, 34, 36, 35,

Table 6.3 Number of Consonants, Vowels, and Silence present in the phone-set considered by Urdu and Assamese Mono-PRSs

Mono-PRS	Count of Different Phonetic units			
	Consonants	Vowels	Silence	Total
Urdu	28	6	1	35
Assamese	24	7	1	32

Table 6.4 Phone error rates of monolingual phone recognition systems

Mono-PRSs	PERs (%)			
	CI		CD	
	HMM	DNN	HMM	DNN
Kannada	43.5	39.5	38.5	37.1
Telugu	42.1	35.5	35.0	30.7
Bengali	49.0	41.6	43.4	37.6
Odia	33.6	29.5	28.0	26.5
Urdu	41.7	36.5	36.6	33.7
Assamese	51.0	46.8	48.4	46.2

and 32 phones, respectively. Mono-PRSs are used in the second stage of LID-Mono systems as shown in Fig. 6.3. The phone-level statistics of KN, TE, BN, and OD Mono-PRSs are same as given in Table 3.2, whereas the phone-level statistics of AS and UR Mono-PRSs are given in Table 6.3. First column indicates the name of Mono-PRSs. Second to fifth columns show the number of consonants, vowels, silence, and the total number of phones, respectively, for each Mono-PRS.

Both HMMs and DNNs are explored for training Mono-PRSs under CI and CD setups. Sclite tool [22] is used for computing the PERs using the procedure mentioned in Sect. 3.4. Table 6.4 shows the PERs of six Mono-PRSs based on MFCC features. Although the PERs of Kannada, Telugu, Bengali, and Odia Mono-PRSs are given in Table 3.4, they are repeated here for convenience and better analysis. First column shows the language used for building the Mono-PRS. Second and third columns provide the PERs of the CI systems using HMMs and DNNs, respectively. Similarly, the last two columns tabulate the results of CD systems. Among all the Mono-PRSs, the Assamese Mono-PRS has shown very poor performance, which is possibly due to the low-resource consideration, noisy transcriptions, tonality, and inadequate occurrences of certain phones in the phone-set. It can be noted that Assamese speech corpus had least amount of speech data of about 2.39 h compared to all other languages (see Tables 3.1 and 6.1). In all the cases, the CD models have lower PERs than their corresponding CI models. It is observed that DNNs have superior performance compared to HMMs. Hence, only CD DNNs are used to develop monolingual or multilingual phone recognizers in all of the remaining experiments.

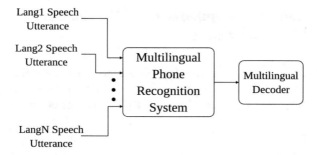

Fig. 6.5 Multilingual Phone Recognition using Common Multilingual Phone-set Approach

6.3.2 Multilingual Phone Recognition using Common Multilingual Phone-set (Multi-PRS) Approach

The Multi-PRS approach is same as the one described in Sect. 3.3.2 of Chap. 3. Figure 6.5 shows the schematic representation of the Multi-PRS. Unlike Fig. 6.3, which has two stages, Fig. 6.5 has a single stage. Multi-PRS, irrespective of the language, can accept the speech input from any language and decode it into a sequence of phonetic units using a common multilingual phone-set [23]. Some of the notable works based on based on *common multilingual phone-set* approach are reported in [24–29] and are briefly described in Sect. 2.2.

The data from six Indian languages—KN, TE, BN, OD, UR, and AS—is used for developing the Multi-PRS. The use of IPA transcription to derive the *common multilingual phone-set* is proposed. Since the IPA has strict one-to-one correspondence between symbols and sounds, the use of IPA can accommodate any sound unit that is present in world's diverse languages. The common multilingual phone-set is derived by grouping the acoustically similar IPAs from all the languages together and selecting the phonetic units which have sufficient number of occurrences to train a separate model for each of them. The IPA symbols that do not have sufficient number of occurrences are mapped to the closest linguistically similar phonetic units present in the common multilingual phone-set. For better analysis and comparison, separate Multi-PRSs are developed using the data of 4 (KN-BN-OD-UR), 5 (KN-TE-BN-OD-UR), and 6 (KN-TE-BN-OD-UR-AS) languages. The number of phones present in the common multilingual phone-set of these 4, 5, and 6 languages is 44, 46, and 46, respectively. For the experiments involving four languages, unlike in previous chapters which use *KN-TE-BN-OD* languages, in this chapter *KN-BN-OD-UR* languages are used. This is because, the code-switching experiments considered in this study use *KN-UR* test sets, and hence we wanted to use KN-UR languages mandatorily for training all our multilingual systems. Hence, *KN-BN-OD-UR* languages are used instead of *KN-TE-BN-OD* for building multilingual systems of 4 languages.

6.4   Evaluation and Comparison of LID-Mono and Multi-PRS Approaches

In this section, the LID-Mono and Multi-PRS approaches for multilingual phone recognition are evaluated and the results are compared using non-code-switched and code-switched test sets. The advantages and limitations of two approaches are discussed.

6.4.1   Non-Code-Switched Scenario

All the experiments related to non-code-switched scenario are conducted using the test set of multilingual speech corpora (see Tables 3.1 and 6.1). The test sets of all the languages involved in training a multilingual phone recognizer become the test set for testing that specific multilingual phone recognizer. The utterances in the test set do not have any code-switching and each utterance belongs to a single language. For comparison and analysis, *Oracle-LID-Mono* system is developed in addition to the LID-Mono and Multi-PRS systems. The *Oracle-LID-Mono* sets the best performance achievable by the LID-Mono approach. A LID-Mono system with LID accuracy equal to 100% is called *Oracle-LID-Mono* system. In *Oracle-LID-Mono* system, the LID block has an ideal LID system without any errors or misclassifications. In order to simulate the *Oracle-LID-Mono* system, the language of each test file present in the test set is manually labeled, followed by decoding it using the monolingual phone recognizer of the same language.

Table 6.5 shows the PERs of multilingual phone recognizers based on LID-Mono and Multi-PRS approaches using non-code-switched test sets. First column indicates the languages used in multilingual phone recognition. Second and third columns show the PERs of Multi-PRS and LID-Mono (using i-vectors) approaches, respectively. Last column shows the PERs of *Oracle-LID-Mono* system.

It is found that the Multi-PRS and *Oracle-LID-Mono* systems outperform the LID-Mono systems in all the cases. The PERs of *Oracle-LID-Mono* and Multi-PRS systems differ by a very small margin. The *Oracle-LID-Mono* systems perform better than Multi-PRS for 4 and 5 languages, whereas for 6 languages Multi-PRS has superior performance compared to *Oracle-LID-Mono* system. The 35.3% PER of *KN-TE-BN-OD-UR-AS Oracle-LID-Mono* system using 6 languages is possibly

Table 6.5 Phone error rates of multilingual phone recognition systems based on LID-Mono and Multi-PRS approaches using Non-code-switched Test set

Languages	Multi-PRS	LID-Mono(i-vectors)	Oracle-LID-Mono
KN-BN-OD-UR	34.0	35.5	33.73
KN-TE-BN-OD-UR	33.4	35.6	33.12
KN-TE-BN-OD-UR-AS	35.2	37.9	35.30

due to the poor performance of Assamese Mono-PRSs. However, the performance of *KN-TE-BN-OD-UR-AS Multi-PRS* system (i.e. 35.2% PER) is good despite the poor performance of Assamese, which is because of the Multi-PRS developed using *common multilingual phone-set* compensates for various issues such as low-resource availability, noisy transcriptions, inadequate occurrences of certain phones in the phone-set, etc.

The absolute error reductions in the PERs of Multi-PRSs compared to LID-Mono systems are 1.5%, 2.2%, and 2.7% for 4, 5, and 6 languages. This clearly indicates that as the number of languages increases Multi-PRSs will have better performance compared to LID-Mono systems. Higher the number of languages, more the benefit from Multi-PRSs compared to LID-Mono systems. As the number of languages increases, the performance of LID-Mono decreases drastically. However, the decrease in the performance of Multi-PRS with the increasing number of languages is not as drastic as that of LID-Mono systems. Since the common multilingual phone-set of Multi-PRS has relatively higher number of phones compared to Mono-PRSs, the use of Multi-PRS has an additional advantage of decoding the speech utterance into relatively higher number of phones compared to the LID-Mono approach. This might help in improving the overall speech recognition accuracy.

Figure 6.6 shows the performance of LID-Mono and Multi-PRSs on test data drawn from 4, 5, and 6 languages in terms of % LID accuracy and % PER. It can be noted that the LID-Mono has an inherently poor performance marked by decreasing % LID accuracy as the number of language classes increase from 4 to 6, which in turn impacts the % PER to increase in going from 4 to 6 languages. When the LID system makes an error, the LID-switched monolingual phone recognition chooses the wrong language phone acoustic models to decode the input speech and naturally incurs a higher PER. The PERs of Multi-PRS, in contrast, have a robust constant performance across the multiple languages clearly arising from its not depending on a front-end LID decision making.

In addition to the better PERs, Multi-PRS approach offers the following advantages over LID-Mono approach:

- Multi-PRS approach does not depend on LID block while the LID-Mono approach heavily depends on the accuracy of the LID block. The failure of LID block will result in phone recognition by an incorrect acoustic model resulting in higher PERs. Particularly, in case of closely related languages, which have significant lexical and structural overlap, where the LID system has high chance of failing, Multi-PRS has huge advantages over LID-Mono.
- The two-staged architecture of LID-Mono is comparatively complex than the single-staged Multi-PRS. This results in higher response time and latency of LID-Mono than Multi-PRS.
- Building a separate monolingual phone recognizer for all the languages is not feasible as the resources required to train a Mono-PRS might not be available in all the languages. Particularly, in case of the under-resourced languages such as Konkani, Kokborok, etc., where sufficient amount of digital data is not available.

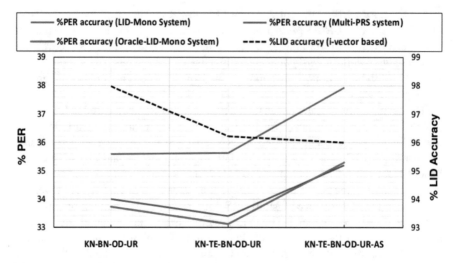

Fig. 6.6 LID accuracy (%) and PERs (%) of LID-Mono and Multi-PRS Approaches using Non-code-switched Test set

6.4.2 Code-Switched Scenario

All the experiments related to code-switched scenario are conducted using the test set described in Sect. 6.2.2. For better comparison and analysis, in addition to the multilingual phone recognizers based on 4, 5, and 6 languages, the bilingual phone recognizer based on KN and UR languages is also developed using both LID-Mono and Multi-PRS approaches. This is because our code-switched test set contained KN and UR languages. The experiments are conducted by varying the length of the segments used to perform LID. The length of the segments is termed as *window size* and represented in terms of seconds. Table 6.6 shows the PERs of LID-Mono systems under varying window sizes using KN-UR code-switched test data. First column shows the window sizes (in seconds). Second to fifth columns provide the PERs of LID-Mono systems based on different language sets. It is observed that the PERs of all LID-Mono systems have consistently increased from small window size to large window size. The least PER is obtained when full speech utterance is used in all LID-Mono systems.

Table 6.7 shows the PERs of multilingual phone recognition systems using code-switching test set. First column shows the languages involved in building the multilingual phone recognizer. Second column shows the PERs of Multi-PRS. The PERs of best performing LID-Mono systems are provided in third column.

It is observed that the Multi-PRSs outperform LID-Mono systems in all the cases. Among four Multi-PRSs, *KN-BN-OD-UR* has least PER while the *KN-UR* has highest PER. The error difference between least and highest PER of LID-Mono systems is 0.3% whereas the similar error difference in case of Multi-PRSs is 1.2%. In case of LID-Mono, PERs get worse as the number of languages increases, which

Table 6.6 PERs of LID-Mono systems under varying window durations using code-switched test set

Size	LID-Mono PERs (%)			
	KN-TE-BN-OD-UR-AS	KN-TE-BN-OD-UR	KN-BN-OD-UR	KN-UR
0.5	61.1	61.0	58.6	58.1
1.0	45.9	45.9	43.9	42.6
1.5	41.6	41.5	40.3	39.7
2.0	39.5	39.4	38.8	38.5
2.5	38.6	38.6	37.9	37.8
3.0	38.0	37.5	36.7	36.6
4.0	36.5	36.5	36.1	36.0
5.0	36.5	37.1	35.8	36.0
Full	34.8	34.8	34.5	34.6

Table 6.7 Phone error rates of multilingual phone recognition systems using code-switched test set

Languages	Approach	
	Multi-PRS	LID-Mono (i-vectors)
KN-UR	32.7	34.6
KN-BN-OD-UR	31.5	34.5
KN-TE-BN-OD-UR	32.5	34.8
KN-TE-BN-OD-UR-AS	31.9	34.8

is mainly due to the poor performance of LID with increasing number of languages. The *KN-BN-OD-UR* Multi-PRS has shown the least PER with an absolute error reduction of 3% (i.e. relative error reduction of 8.69%) compared to *KN-BN-OD-UR* LID-Mono system. The average PER of Multi-PRS is 32.15% whereas the average PER of LID-Mono is 34.67%. The error reduction in the PER of an average Multi-PRS is 2.52% (absolute) compared to the PER of an average LID-Mono system.

Figure 6.7 compares the performance of LID-Mono and Multi-PRS approaches using code-switched test set. It provides a composite display of the PERs of LID-Mono and Multi-PRS for different number of test languages and different window sizes over which the LID makes a decision. The X-axis denotes window sizes varying between 500 ms–5 s and the full utterance, and the Y-axis denotes PERs of multilingual phone recognizers.

All the curves denoting Multi-PRS are at the bottom. There is a clear gap between the curves denoting Multi-PRSs and the curves denoting LID-Mono systems indicating that all the Multi-PRSs have superior performance compared to LID-Mono systems. There is clear reduction in PERs of LID-Mono systems as the size of window increases. For the full utterance case, all the curves of LID-Mono systems have almost converged to a single point indicating that the performance of all the LID-Mono systems are very similar to each other. On the contrary, all the curves of Multi-PRSs are straight lines, this is because there is no need of windowing in Multi-PRS as it does not have a LID system. All the Multi-PRSs offer a consistent performance that is superior compared to any of the

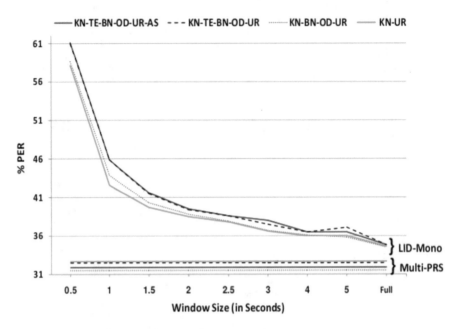

Fig. 6.7 Comparison of LID-Mono and Multi-PRS Approaches using Code-switched Test set

LID-Mono systems, independent of window size, and hence is robust to arbitrary code-switching durational distributions.

It can be noted that the LID-Mono systems assume that all the windows in a code-switched speech utterance are independent of each other and can be easily separated. However, this is not true in reality as multiple code-switches can happen within a window. The problem of language boundary detection is not taken into account in LID-Mono approach. The use of accurate time information at which the code-switching happens (from one language to another) as an additional knowledge source during decoding can improve the performance of LID-Mono systems. However, the accurate detection of language boundaries in a code-switched speech data has its own challenges. This makes the Multi-PRS a natural choice to recognize code-switched speech, with practically no particular merit to choose the LID-Mono system, which suffers from the trade-off discussed above, higher design complexity of having to design a LID system (to recognize multiple language classes), having to possibly detect language switch boundaries and having to design multiple monolingual phone recognition systems.

6.5 Summary

The LID-Mono and Multi-PRS approaches of multilingual phone recognition are evaluated and compared. The i-vectors have shown better LID accuracy compared to MFCCs. The performance characteristics of LID-Mono and Multi-PRS approaches are studied with respect to several underlying parameters such as the interval over which the LID makes a decision, the number of languages on which the LID is designed for the LID-Mono systems, and the means of arriving at a common phone-set for the Multi-PRS. It is found that the performance of Multi-PRS approach is superior compared to the more conventional LID-Mono approach in both non-code-switched and code-switched scenarios. For code-switched speech, the effect of length of segments (that are used to perform LID) on the performance of LID-Mono system is studied by varying the window size from 500 ms to 5 s, and full utterance. The LID-Mono approach heavily depends on the accuracy of the LID system and the LID errors cannot be recovered. But, the *common multilingual phone-set* based Multi-PRS by virtue of not having to do a front-end LID switching and designed based on the common multilingual phone-set derived from several languages, it is not constrained by the accuracy of the LID system and hence performs effectively on non-code-switched and code-switched speech offering low PERs than the LID-Mono system.

References

1. C. Chang, C. Lin, LIBSVM: A library for support vector machines. ACM Trans. Intell. Syst. Technol. **2**(3), 1–27 (2011). https://doi.org/10.1145/1961189.1961199
2. K.E. Manjunath, S.S. Kumar, D. Pati, B. Satapathy, K.S. Rao, Development of consonant-vowel recognition systems for Indian languages: Bengali and Odia, in *IEEE India Conference on Emerging Trends and Innovation in Technology (INDICON)* (2013), pp. 1–6. https://doi.org/10.1109/INDCON.2013.6726109
3. scikit-learn, Scikit-learn: machine learning in Python—Online documentation. https://scikit-learn.org [Accessed Mar. 08, 2020]
4. N. Dehak, P.A. Torres-Carrasquillo, D. Reynolds, R. Dehak, Language recognition via i-vectors and dimensionality reduction, in *INTERSPEECH* (2011), pp. 857–860
5. D.A. Reynolds, T.F. Quatieri, R.B. Dunn, Speaker verification using adapted Gaussian mixture models, in *Digital Signal Processing* (2000), pp. 19–41
6. B. Jiang, Y. Song, S. Wei, J.H. Liu, I. McLoughlin, L. Dai, Deep bottleneck features for spoken language identification. PLoS ONE **9**(4), 1–11 (2014). https://doi.org/10.1371/journal.pone.0100795
7. B. Jiang, Y. Song, S. Wei, M. Wang, I. McLoughlin, L. Dai, Performance evaluation of deep bottleneck features for spoken language identification, in *International Symposium on Chinese Spoken Language Processing* (2014), pp. 143–147. https://doi.org/10.1109/ISCSLP.2014.6936580
8. B. Padi, S. Ramoji, V. Yeruva, S. Kumar, S. Ganapathy, The LEAP language recognition system for LRE 2017 challenge—improvements and error analysis, in *Odyssey: The Speaker and Language Recognition Workshop* (2018), pp. 31–38. https://doi.org/10.21437/Odyssey.2018-5

9. K.E. Manjunath, K.M.S. Raghavan, K.S. Rao, D.B. Jayagopi, V. Ramasubramanian, Multilingual phone recognition: comparison of traditional versus common multilingual phone-set approaches and applications in code-switching, in *International Symposium on Signal Processing and Intelligent Recognition Systems*, Thiruvananthapuram (2019). https://doi.org/10.1007/978-981-15-4828-4_7

10. K.E. Manjunath, K.M.S. Raghavan, K.S. Rao, D.B. Jayagopi, V. Ramasubramanian, Approaches for multilingual phone recognition in code-switched and non-code-switched scenarios using Indian languages, in *ACM Transactions on Asian and Low-Resource Language Information Processing (TALLIP)* (2020)

11. A. Waibel, H. Soltau, T. Schultz, T. Schaaf, F. Metze, Multilingual Speech Recognition, in *Verbmobil: Foundations of Speech-to-Speech Translation. Artificial Intelligence* (Springer, Berlin, 2000), pp. 33–45. https://doi.org/10.1007/978-3-662-04230-4_3

12. H. Lin, J.T. Huang, F. Beaufays, B. Strope, Y. Sung, Recognition of multilingual speech in mobile applications, in *IEEE International Conference on Acoustics, Speech, and Signal Processing (ICASSP)*, Kyoto (2012), pp. 4881–4884. https://doi.org/10.1109/ICASSP.2012.6289013

13. J.G. Dominguez, D. Eustis, I.L. Moreno, A. Senior, F. Beaufays, P.J. Moreno, A real-time end-to-end multilingual speech recognition architecture. IEEE J. Sel. Top. Signal Process. **9**(4), 749–759 (2015). https://doi.org/10.1109/JSTSP.2014.2364559

14. A.K.V. SaiJayram, V. Ramasubramanian, T.V. Sreenivas, Language identification using parallel sub-word recognition, in *IEEE International Conference on Acoustics, Speech, and Signal Processing (ICASSAP)* (2003), pp. I–32. https://doi.org/10.1109/ICASSP.2003.1198709

15. S.A. Santosh Kumar, V. Ramasubramanian, Automatic language identification using ergodic-hmm, in *ICASSP* (2005), pp. 609–612. https://doi.org/10.1109/ICASSP.2005.1415187

16. L. Mary, B. Yegnanarayana, Autoassociative neural network models for language identification, in *International Conference on Intelligent Sensing and Information Processing* (2004), pp. 317–320. https://doi.org/10.1109/ICISIP.2004.1287674

17. D. Nandi, D. Pati, K.S. Rao, Implicit processing of LP residual for language identification. Comput. Speech Lang. **41**(C), 68–87 (2017). https://doi.org/10.1016/j.csl.2016.06.002

18. T. Nagarajan, H.A. Murthy, A pair-wise multiple codebook approach to implicit language identification, in *Workshop on Spoken Language Processing* (2003), pp. 101–108

19. M. Li, H. Suo, X. Wu, P. Lu, Y. Yan, Spoken language identification using score vector modeling and support vector machine, in *INTERSPEECH* (2007), pp. 350–353

20. W.M. Campbell, E. Singer, P.A. Torres-Carrasquillo, D.A. Reynolds, Language recognition with support vector machines, in *Proceedings of the Odyssey: The Speaker and Language Recognition Workshop* (2004), pp. 285–288

21. W.M. Campbell, J.P. Campbell, D.A. Reynolds, E. Singer, P.A. Torres-Carrasquillo, Support vector machines for speaker and language recognition. Comput. Speech Language **20**(2–3), 210–229 (2006). https://doi.org/10.1016/j.csl.2005.06.003

22. Sclite Tool. http://www1.icsi.berkeley.edu/Speech/docs/sctk-1.2/sclite.htm [Accessed Mar. 08, 2020]

23. S.M. Siniscalchi, D. Lyu, T. Svendsen, C. Lee, Experiments on cross-language attribute detection and phone recognition with minimal target-specific training data. IEEE Trans. Acoust. Speech Signal Process. **20**(3), 875–887 (2012). https://doi.org/10.1109/TASL.2011.2167610

24. T. Schultz, A. Waibel, Language independent and language adaptive acoustic modeling for speech recognition. Speech Commun. **35**, 31–51 (2001). https://doi.org/10.1016/S0167-6393(00)00094-7

25. T. Schultz, A. Waibel, Language independent and language adaptive large vocabulary speech recognition, in *International Conference on Spoken Language Processing (ICSLP)* (1998), pp. 1819–1822

26. T. Schultz, A. Waibel, Multilingual and crosslingual speech recognition, in *Proceedings of the DARPA Workshop on Broadcast News Transcription and Understanding* (1998), pp. 259–262

27. T. Schultz, K. Kirchhoff, *Multilingual Speech Processing* (Academic Press, New York, 2006). https://doi.org/10.1016/B978-0-12-088501-5.X5000-8
28. C.S. Kumar, V.P. Mohandas, L. Haizhou, Multilingual speech recognition: a unified approach, in *INTERSPEECH* (2005)
29. N.T. Vu, D. Imseng, D. Povey, P. Motlicek, T. Schultz, H. Bourlard, Multilingual deep neural network based acoustic modeling for rapid language adaptation, in *IEEE International Conference on Acoustics, Speech, and Signal Processing (ICASSP)*, Florence (2014), pp. 7639–7643. https://doi.org/10.1109/ICASSP.2014.6855086

Chapter 7
Summary and Conclusion

7.1 Summary of the Book

In this work, various aspects of multilingual phone recognition such as development, analysis, performance improvement, and applications are studied using six Indian languages—Kannada, Telugu, Bengali, Odia, Urdu, and Assamese. The IPA based transcription is used for deriving the *common multilingual phone-set* by grouping the acoustically similar phonetic units from multiple languages. The behaviour of Multi-PRS across *two* language families—Dravidian and Indo-Aryan—is compared and analysed by developing separate Multi-PRSs for Dravidian and Indo-Aryan language families. The performance of Multi-PRS is analysed and compared with that of the Mono-PRSs. Both DNNs and HMMs are explored for training PRSs, and it is shown that DNNs outperform HMMs. The context dependent systems have superior performance compared to context independent systems. It is found that the long vowels are better modelled by Dravidian PRSs, while the Indo-Aryan PRSs model the aspirated consonants more accurately. The performance of Multi-PRSs is improved using tandem features.

In this study, the AFs are explored to improve the performance of Multi-PRS. AF-predictors are trained using DNNs for predicting the AFs from MFCCs. Five AF groups—namely place, manner, roundness, frontness, and height—are considered. The oracle AFs are used to set the best performance realizable by the predicted AFs. The performance of predicted and oracle AFs is compared. Two approaches for fusion of AFs, namely LRA and AF-Tandem, are explored and found that AF-Tandem method outperforms LRA method. The MTL is explored to improve the prediction accuracy of AFs, and it is found that the MTL based AF-predictors outperform non-MTL based AF-predictors. The use of AFs has improved the performance of Multi-PRSs. In addition to the AFs, the phone posteriors are fused to further boost the performance of Multi-PRS. The fusion of *MTL-2* based AFs and phone posteriors using tandem approach has resulted in least PER. The best performing system using predicted AFs has shown 3.2% reduction in absolute PER

Manjunath K. E., *Multilingual Phone Recognition in Indian Languages*,
SpringerBriefs in Speech Technology, https://doi.org/10.1007/978-3-030-80741-2_7

(9.1% reduction in relative PER) compared to baseline Multi-PRS. The oracle AFs by feature fusion with MFCCs offer a remarkably low target PER of 10.4%, which is 24.7% absolute reduction compared to baseline Multi-PRS with MFCCs alone. Given the remarkably low PER of 10.4% for oracle based Multi-PRS, it is concluded that there is much scope for enhanced prediction of AFs to improve the Multi-PRS to reach the performance of oracle AFs.

In this work, LID-switched monolingual phone recognition (i.e. LID-Mono approach) and *common multilingual phone-set* based approaches for multilingual phone recognition are evaluated and compared using non-code-switched and code-switched test data. The bilingual code-switching experiments are conducted using the code-switched data of Kannada and Urdu languages. MFCCs and i-vectors are explored for performing LID and found that i-vectors have better LID accuracy compared to MFCCs. The performance characteristics of LID-Mono and *common multilingual phone-set* based approaches are studied with respect to several underlying parameters such as the interval over which the LID makes a decision, the number of languages on which the LID is designed for the LID-Mono systems, and the means of arriving at a common multilingual phone-set. It is found that the performance of *common multilingual phone-set* based approach is superior compared to the more conventional LID-Mono approach in both non-code-switched and code-switched scenarios. For code-switched speech, the effect of length of segments (that are used to perform LID) on the performance of LID-Mono system is studied. The LID-Mono approach heavily depends on the accuracy of the LID system and the LID errors cannot be recovered. But the *common multilingual phone-set* based system by virtue of not having to do a front-end LID switching and designed based on the common multilingual phone-set derived from several languages, it is not constrained by the accuracy of the LID system and hence performs effectively on non-code-switched and code-switched speech offering low PERs than the LID-Mono system.

7.2 Contributions of the Book

The major contributions of this book can be summarized as follows:

- Multi-PRS is developed using the *common multilingual phone-set* derived from the IPA based transcription of six Indian languages—Kannada, Telugu, Bengali, Odia, Urdu, and Assamese.
- The performance of Multi-PRS is improved using tandem features. The Mono-PRSs *versus* Multi-PRS and baseline *versus* tandem systems are compared and their results are analysed.
- Methods are proposed to predict articulatory features from spectral features using DNNs. Multitask learning is explored to improve the prediction accuracy of AFs.

- The articulatory features are explored to improve the performance of Multi-PRS using lattice rescoring method of combination and tandem method of combination.
- The LID-Mono and *common multilingual phone-set based* multilingual phone recognition systems are developed and evaluated. The results are compared and analysed in detail.
- The effectiveness and superiority of the proposed *common multilingual phone-set* based system over the LID-Mono approach are demonstrated using non-code-switched and code-switched test data.

7.3   Future Scope of Work

- The size of speech data used in this study varies between 3.8 and 5.7 h for different languages. The current size of the multilingual speech corpora can be increased and similar experiments can be conducted to get better performance. With the increased training data, the DNNs are expected to perform much better.
- This work is primarily focused on phone recognition and reports the performance in terms of PERs. This study does not use any language model. In future, one can employ the language model and measure the performance in terms of WER. This enables to cross-compare with similar systems developed using other languages.
- In this work, two major language families (Dravidian and Indo-Aryan) of Indian languages are studied and compared. Similar studies can be extended to other language families such as Sino-Tibetan, Austroasiatic, etc.
- While doing comparison between the Multi-PRSs of Dravidian and Indo-Aryan language families, only two languages from each language family are considered in this study. This study can be further enhanced by using more than two languages from each language family.
- In this study, only four languages are considered to conduct the experiments related to use of AFs to improve the performance of Multi-PRSs. The number of languages can be increased and the improvements in the performance of Multi-PRSs can be examined.
- It is found that the oracle based Multi-PRS have achieved a remarkably low PER of 10.4%. This indicates that there is much scope for enhanced prediction of AFs so as to improve the performance of Multi-PRS to reach that of oracle AFs. In future, one can explore other alternative methods (including the continuous valued AFs) to improve the accuracy of prediction of AFs.
- One can extend the multilingual framework to train the phone recognizers for low-resource languages (i.e. language adaptation) and to perform language identification.
- In this book, bilingual code-switching scenario is considered to demonstrate the applications of Multi-PRS in code-switching using Kannada and Urdu languages. In future, the code-switching between more than two languages can be studied.

One can also explore the code-switching between other language pairs such as Kannada-Telugu or English-Hindi or Kannada-Hindi, etc.

- This study uses DNN-HMMs to train Multi-PRSs. Other state-of-the-art automatic speech recognition paradigms such as end-to-end speech recognition using recurrent neural networks or convolutional neural networks can be explored.
- In this study, read speech is considered for developing Multi-PRSs. The same framework can be extended to other modes of speech such extempore and conversational modes of speech [1].
- In this work, the problem of language boundary detection in code-switched speech is not taken into account while developing LID-Mono systems. In future, one can incorporate the language boundary information into LID-Mono approach and improve the performance of LID-Mono systems.

Reference

1. S.B.S. Kumar, K.S. Rao, D. Pati, Phonetic and prosodically rich transcribed speech corpus in Indian languages: Bengali and Odia, in *Proceedings of the Sixteenth IEEE International Oriental COCOSDA*, Gurgaon (2013), pp. 1–5. https://doi.org/10.1109/ICSDA.2013.6709901

Appendix A
Support Vector Machines

Support Vector Machines (SVMs) are supervised learning models associated with learning algorithms that analyse data and recognize patterns. SVMs are used for classification and regression analysis. The basic SVM takes a set of input data and predicts, for each given input, which of two possible classes forms the output. This SVM is called binary linear classifier. In the feature space, an optimal hyperplane discrimination function $y = \mathbf{w}\mathbf{x} + b$ is constructed to separate samples from two classes. Let $\mathbf{x_i}$ be the training vectors with $i = 1, 2, .., l$ in two classes. The desired output y_i for the training example x_i is defined as follows:

$$y_i = \begin{cases} +1 \text{ if } \mathbf{x_i} \in \text{ class } 1 \\ -1 \text{ if } \mathbf{x_i} \in \text{ class } 2. \end{cases}$$

The examples with $y_i = +1$ are called positive examples, and those with $y_i = -1$ are negative ones. An optimal hyperplane is constructed to separate positive examples from negative ones. The separating hyperplane (margin) is chosen in such a way so as to maximize its distance from the closest training examples of different classes. Figure A.1 illustrates the geometric construction of hyperplane for two dimensional input space. The support vectors are those data points that lie closest to the decision surface, and therefore the most difficult to classify. They have a direct bearing on the optimum location of the decision surface. Figure A.1 represents a binary classifier.

In addition to performing linear classification, SVMs can efficiently perform a non-linear classification using the kernel trick. The kernel trick is used to implicitly map the inputs into high-dimensional feature spaces. Multi-class SVMs can be constructed in several approaches including one-against-one, one-against-all. One-Against-One also called Max-win voting. This strategy builds one SVM for each pair of classes. In this approach, if N is the number of classes, then $N(N-1)/2$ classifiers are constructed and each one trains data from two classes. The test data is applied to all machines, each classification gives one vote to the winning class,

© The Author(s), under exclusive license to Springer Nature Switzerland AG 2022
Manjunath K. E., *Multilingual Phone Recognition in Indian Languages*,
SpringerBriefs in Speech Technology, https://doi.org/10.1007/978-3-030-80741-2

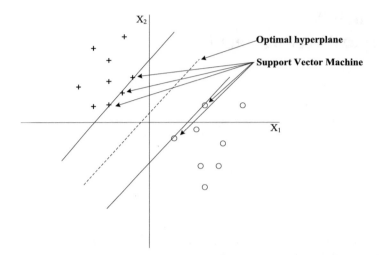

Fig. A.1 Illustration of Support Vector Machines

and the data is labeled with the class having most votes. If there are two identical votes, Max-win selects the class with the smallest index. The radial basis function can be used as a kernel. One-against-all strategy involves training a single classifier per class, with the samples of that class as positive samples and all other samples as negatives. This strategy requires the base classifiers to produce a real-valued confidence score for its decision, rather than just a class label. This is because, discrete class labels alone can lead to ambiguities, where multiple classes are predicted for a single sample.

Appendix B
Hidden Markov Models for Speech Recognition

Hidden Markov Models (HMMs) are commonly used classification models in speech recognition [1]. HMMs are used to capture the sequential information present in feature vectors for developing speech recognizers. HMM is a stochastic signal model which is referred to as Markov sources or probabilistic functions of Markov chains. This model is an extension to the concept of Markov model which includes the case where the observation is a probabilistic function of the state. HMM is a finite set of states, each of which is associated with a probability distribution. Transitions among the states are governed by a set of probabilities called transition probabilities. In a particular state an outcome or an observation can be generated, according to the associated probability distribution. Only the outcome is known and underlaying state sequence is hidden. Hence, it is called a hidden Markov model.

The following are the five basic elements that define HMM

1. N, the number of states in the model,
   $s = \{1, 2, \ldots i, \ldots j, \ldots, N\}$.
2. M, number of distinct observation symbols per state,
   $v = \{v_1, v_2, \ldots, v_M\}$.
3. State transition probability distribution $A = \{a_{ij}\}$ where

$$a_{ij} = P[q_{t+1} = j | q_t = i], 1 \leq i, j \leq N. \tag{B.1}$$

4. Observation symbol probability distribution in state j,
   $B = \{b_j(k)\}$ where

$$b_j(k) = P[v_k \text{ at } t | q_t = j] \qquad 1 \leq j \leq N, 1 \leq k \leq M \tag{B.2}$$

Initial state distribution $\pi = \{\pi_j\}$ where

$$\pi_j = P[q_1 = j] \qquad 1 \leq j \leq N. \tag{B.3}$$

© The Author(s), under exclusive license to Springer Nature Switzerland AG 2022
Manjunath K. E., *Multilingual Phone Recognition in Indian Languages*,
SpringerBriefs in Speech Technology, https://doi.org/10.1007/978-3-030-80741-2

A complete specification of an HMM requires specification of two model parameters (N and M), specification of observation symbols, and the specification of three probability measures A, B, π. HMM is indicated by the compact notation

$$\lambda = (A, B, \pi).$$

Given that state sequence $q = (q_1, q_2, \ldots, q_T)$ is unknown, the probability of observation sequence $O = (o_1, o_2, \ldots, o_T)$ given the model λ is obtained by summing the probability of over all possible state sequences q as follows:

$$P(O|\lambda) = \sum_{q_1, q_2, \ldots, q_T} \pi_{q_1} b_{q_1}(o_1) a_{q_1 q_2} b_{q_2}(o_2) \ldots a_{q_{T-1} q_T} b_{q_T}(o_T), \qquad (B.4)$$

where π_{q_1} is the initial state probability of q_1 and T is length of the observation sequence.

Reference

1. L.R. Rabiner, A tutorial on hidden Markov models and selected applications in speech recognition, in *Proceedings of IEEE* (1989), pp. 257–286.

Appendix C
Deep Neural Networks for Speech Recognition

A Deep Neural Network (DNN) is a feedforward, artificial neural network that has more than one layer of hidden units between its inputs and its outputs. An artificial neural network is a computational model that is loosely inspired by the human brain consisting of an interconnected network of simple processing units that can learn from experience by modifying its connections [1]. The training of an Artificial Neural Network system can be of two types. First, supervised learning where a set of labeled input data is used for trainings. The labels are the desired outputs. On the other hand, in unsupervised learning, the network learns the structure of the given input and learns through feedback for the outcome it produces.

C.1 FeedForward Neural Networks

The artificial neural networks in which the information moves from the input layer to output layer through the hidden layer in forward direction with no loops in the network are called FeedForward Neural Networks (FFNNs). FFNNs are used to capture the non-linear relationship between the feature vectors and the phonetic sound units. Each unit in one layer of the FFNN has directed connections to the units in the subsequent layer. FFNNs consist of an input layer, an output layer, and one or more hidden layers. In case of phone recognition, FFNNs are used to map an input feature vector into one of the phonetic unit among the set of phonetic units used for training the FFNN models. The number of units in the input is equal to the dimension of feature vectors whereas the number of units in output layer is equal to number of phonetic sound units being modelled. Figure C.1 shows the structure of a simple three layered FFNNs. A three layered FFNN has one input layer, one hidden layer, and one output layer.

The hidden and output layers are non-linear, whereas the input layer is linear. The non-linearity is achieved using an activation function. An activation function

© The Author(s), under exclusive license to Springer Nature Switzerland AG 2022   93
Manjunath K. E., *Multilingual Phone Recognition in Indian Languages*,
SpringerBriefs in Speech Technology, https://doi.org/10.1007/978-3-030-80741-2

Fig. C.1 General structure of
three layered FeedForward
Neural Networks

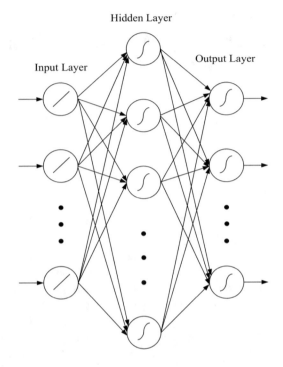

determines the relation between the input and output. Activation function normalizes the output value to a certain range. There are a number of activation functions that are used with neural networks. Some of the widely used activation functions are sigmoid, ReLU, McCulloch-Pitts thresholding function, and piecewise-linear function. Sigmoid function has s-shaped graph and is most common form of activation function for defining the output of a neuron. Sigmoid function is strictly increasing, continuous, and differential function. The sigmoid function is a widely used function for FFNNs with back-propagation because of its non-linearity and simplicity of computation [2]. An example of the sigmoid function is the logistic function given by Eq. (C.1).

$$\phi(v) = \frac{1}{1 + \exp(-av)}, \tag{C.1}$$

where a is the slope parameter of the sigmoid function and v is the local output of the neuron. Hyperbolic tangent sigmoid activation function is symmetric bipolar activation function given by Eq. (C.2).

$$\phi(v) = \frac{2}{(1 + \exp(-2v))} - 1. \tag{C.2}$$

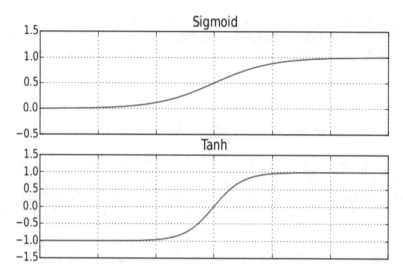

Fig. C.2 Sigmoid and Hyperbolic Tangent Activation functions used in DNN studies

Figure C.2 shows the sigmoid and hyperbolic tangent activation functions that are commonly used in DNN studies.

The output layer of the neural network outputs the posterior probability predictions for a realized model. The output values from the units of the last layer transform values into probabilities via a softmax function. The feature vectors are fed to the input layer and the corresponding phone labels are fed to the output layer of the FFNN. FFNNs are trained using a learning algorithm such as backpropagation algorithm [3, 4]. The back-propagation algorithm is most commonly used in the development of speech recognition applications.

Back-propagation is the process of minimizing the differences between the actual output and the desired output or the error based on the training samples with labels. The target is to minimize the error. An error occurs when an input does not produce the desired output. In supervised learning, the system is trained with data where for each input the correct output is known. A predefined error-function is used to compute the error between the actual and desired outputs. The error is then back propagated through the network and the weights of the network are adjusted based on the computed error. The weights are adjusted using a non-linear optimization method such as stochastic gradient descent, limited memory BFGS, conjugate gradient, etc. In the optimization process, the weights are varied to minimize the error based on the given samples during the training. It is also necessary that the optimization does not over-fit to the training data and can generalize well for unseen test data. The objective of an optimizer is to get to the minimum point of the error curve for different weights. This process is repeated for sufficiently large number of training examples till the network converges. After the completion of training phase, the weights of the network are used for decoding the phonetic sound units in the spoken utterances. Determining the network structure is again an optimization problem. The key factors that influence the neural network structure are amount of training data, learning ability of the network, and capacity to generalize the acquired knowledge.

C.2   Training Deep Neural Networks

A DNN is a feedforward, artificial neural network that has more than one layer of hidden units between its inputs and its outputs. Each hidden unit, j, typically uses the logistic function or a closely related hyperbolic tangent function to map its total input from the layer below x_j, to the scalar state, y_j that it sends to the layer above.

$$y_j = logistic(x_j) = \frac{1}{1 + e^{-x_j}}, x_j = b_j + \sum_i y_i w_{ij}, \qquad (C.3)$$

where b_j is the bias of unit j, i is an index over units in the layer below, and w_{ij} is the weight on a connection to unit j from unit i in the layer below. For multi-class classification, output unit j converts its total input, x_j, into a class probability, p_j, by using the *softmax* non-linearity:

$$p_j = \frac{exp(x_j)}{\sum_k exp(x_k)}, \qquad (C.4)$$

where k is an index over all classes.

DNNs can be discriminatively trained by back-propagating derivatives of a cost function that measures the discrepancy between the target outputs and the actual outputs produced for each training case [5]. When using the softmax output function, the natural cost function C is the cross-entropy between the target probabilities d and the outputs of the softmax, p:

$$C = -\sum_j d_j log p_j, \qquad (C.5)$$

where the target probabilities, typically taking values of one or zero, are the supervised information provided to train the DNN classifier [6].

C.3   Interfacing DNN with HMM (DNN-HMMs)

Once a DNN has been discriminatively fine-tuned, it outputs probabilities of the form $p(HMMstate|AcousticInput)$. But to compute a Viterbi alignment or to run the forward-backward algorithm within the HMM framework we require the likelihood $p(AcousticInput|HMMstate)$. The posterior probabilities that the DNN outputs can be converted into the scaled likelihood by dividing them by the frequencies of the HMM-states in the forced alignment that is used for fine-tuning the DNN [7]. All of the likelihoods produced in this way are scaled by the same unknown factor of $p(AcousticInput)$, but this has no effect on the alignment. Although this conversion appears to have little effect on some recognition tasks, it can be important for tasks where training labels are highly unbalanced (e.g. with many frames of silences). Detailed description about the use of DNNs for speech recognition can be found in [6, 8, 9].

References

1. M.V. Gerven, Computational Foundations of Natural Intelligence, in *Frontiers in Computational Neuroscience*, vol. 11 (2017)
2. D. Ponce-Morado, DNN-based Phoneme Models for Speech Recognition, Master's thesis (Swiss Federal Institute of Technology, Zurich, 2015)
3. R. Rojas, *Neural Networks—A Systematic Introduction* (Springer, New-York, 1996)
4. M. Nielsen, *Neural Networks and Deep Learning*. http://neuralnetworksanddeeplearning.com
5. D.E. Rumelhart, G.E. Hinton, R.J. Williams, Learning representations by back-propagating errors. Nature **323**, 533–536 (1986)
6. G. Hinton, L. Deng, D. Yu, G. Dahl, A. Mohamed, N. Jaitly, A. Senior, V. Vanhoucke, P. Nguyen, T. Sainath, B. Kingsbury, Deep neural networks for acoustic modeling in speech recognition, in *IEEE Signal ProcessingMagazine*, vol. 2 (2012)
7. H. Bourlard, N. Morgan, *Connectionist Speech Recognition: A Hybrid Approach* (Kluwer Academic Publishers, Norwell, 1993)
8. L. Deng, D. Yu, Deep learning: methods and applications, in *Foundations and Trends in Signal Processing* (2014)
9. L. Deng, Y. Liu, *Deep Learning in Natural Language Processing* (Springer, Berlin, 2018)

Index

© The Author(s), under exclusive license to Springer Nature Switzerland AG 2022
Manjunath K. E., *Multilingual Phone Recognition in Indian Languages*,
SpringerBriefs in Speech Technology, https://doi.org/10.1007/978-3-030-80741-2

Printed in the United States
by Baker & Taylor Publisher Services